45 mit 620
lbj 27203

Ausgeschieden im Jahr 2025

# The Field Orientation Principle in Control of Induction Motors

# THE KLUWER INTERNATIONAL SERIES IN ENGINEERING AND COMPUTER SCIENCE

## POWER ELECTRONICS AND POWER SYSTEMS

*Consulting Editor*

**Thomas A. Lipo**
University of Wisconsin-Madison

Other books in the series:

SPOT PRICING OF ELECTRICITY, F. C. Schweppe, M. C. Caramanis, R. D. Tabors, R. E. Bohn
  ISBN 0-89838-260-2

RELIABILITY ASSESSMENT OF LARGE ELECTRIC POWER SYSTEMS, R. Billington, R. N. Allan
  ISBN 0-89838-266-1

MODERN POWER SYSTEMS CONTROL AND OPERATION, A. S. Debs
  ISBN: 0-89838-265-3

ELECTROMAGNETIC MODELLING OF POWER ELECTRONIC CONVERTERS, J. A. Ferreira
  ISBN: 0-7923-9034-2

ENERGY FUNCTION ANALYSIS FOR POWER SYSTEM STABILITY, M. A. Pai
  ISBN: 0-7923-9035-0

INDUSTRIAL ENERGY MANAGEMENT: PRINCIPLES AND APPLICATIONS, G. Petrecca
  ISBN: 0-7923-9305-8

# The Field Orientation Principle in Control of Induction Motors

Andrzej M. Trzynadlowski
University of Nevada, Reno

KLUWER ACADEMIC PUBLISHERS
BOSTON/DORDRECHT/LONDON

**Distributors for North America:**
Kluwer Academic Publishers
101 Philip Drive
Assinippi Park
Norwell, Massachusetts 02061 USA

**Distributors for all other countries:**
Kluwer Academic Publishers Group
Distribution Centre
Post Office Box 322
3300 AH Dordrecht, THE NETHERLANDS

**Library of Congress Cataloging-in-Publication Data**

Trzynadlowski, Andrzej.
    The field orientation principle in control of induction motors / Andrzej M. Trzynadlowski.
        p.   cm. -- (The Kluwer international series in engineering and computer science : SECS 258. Power electronics and power systems)
    Includes bibliographical references (p.  ) and index.
    ISBN 0-7923-9420-8 (alk. paper)
    1. Electric motors, Induction--Automatic control.  2. Field orientation principle (Electric engineering)   I. Title.
II. Series: Kluwer international series in engineering and computer science : SECS 258.   III. Series: Kluwer international series in engineering and computer science. Power electronics & power systems.
TK2785.T76   1994
621.46--dc20                                                 93-33661
                                                                                   CIP

Copyright © 1994 by Kluwer Academic Publishers

All rights reserved. No part of this publication may be reproduced, stored in a retrieval system or transmitted in any form or by any means, mechanical, photo-copying, recording, or otherwise, without the prior written permission of the publisher, Kluwer Academic Publishers, 101 Philip Drive, Assinippi Park, Norwell, Massachusetts 02061.

*Printed on acid-free paper.*

Printed in the United States of America

*To the memory of my Grandmother*

*Izabela Prószyńska*

# Contents

| | | |
|---|---|---|
| Nomenclature | | ix |
| Preface | | xv |
| 1 | DYNAMIC MODEL OF THE INDUCTION MOTOR | 1 |
| | 1.1 Space Vectors in Stator Reference Frame | 1 |
| | 1.2 Direct and Inverse Three-Phase to Stator Reference Frame Transformations | 8 |
| | 1.3 Voltage and Current Equations in Stator Reference Frame | 11 |
| | 1.4 Torque Equation | 16 |
| | 1.5 Dynamic Equivalent Circuit | 20 |
| | 1.6 Direct and Inverse Stator to Excitation Reference Frame Transformations | 25 |
| | 1.7 Motor Equations in Excitation Reference Frame | 28 |
| | 1.8 Examples and Simulations | 30 |
| 2 | SCALAR CONTROL OF INDUCTION MOTORS | 43 |
| | 2.1 The $\Gamma$ Equivalent Circuit of an Induction Motor | 44 |
| | 2.2 Principles of the Constant Volts/Hertz Control | 47 |
| | 2.3 Scalar Speed Control System | 52 |
| | 2.4 The $\Gamma'$ Equivalent Circuit of an Induction Motor | 54 |
| | 2.5 Principles of the Torque Control | 56 |
| | 2.6 Scalar Torque Control System | 59 |
| | 2.7 Examples and Simulations | 66 |
| 3 | FIELD ORIENTATION PRINCIPLE | 87 |
| | 3.1 Optimal Torque Production Conditions | 88 |
| | 3.2 Dynamic Block Diagram of an Induction Motor in the Excitation Reference Frame | 90 |
| | 3.3 Field Orientation Conditions | 93 |
| 4 | CLASSIC FIELD ORIENTATION SCHEMES | 97 |
| | 4.1 Field Orientation with Respect to the Rotor Flux Vector | 98 |
| | 4.2 Direct Rotor Flux Orientation Scheme | 100 |
| | 4.3 Indirect Rotor Flux Orientation Scheme | 106 |

CONTENTS

  4.4 Examples and Simulations  108

5 INVERTERS  125

  5.1 Voltage Source Inverter  126
  5.2 Voltage Control in Voltage Source Inverters  130
  5.3 Current Control in Voltage Source Inverters  138
  5.4 Current Source Inverter  141
  5.5 Examples and Simulations  144

6 REVIEW OF VECTOR CONTROL SYSTEMS  159

  6.1 Systems with Stator Flux Orientation  160
  6.2 Systems with Airgap Flux Orientation  168
  6.3 Systems with Current Source Inverters  175
  6.4 Observers for Vector Control Systems  176
  6.5 Adaptive Schemes  185
  6.6 Position and Speed Control of Field-Oriented Induction Motors  189
  6.7 Examples and Simulations  197

Bibliography  225

Index  253

# Nomenclature

## Principal Symbols

| | |
|---|---|
| $a, b, c$ | switching variables of an inverter |
| $d_X, d_Y, d_Z$ | duty ratios of states $X, Y, Z$ of an inverter |
| $\mathcal{F}_{as}, \mathcal{F}_{bs}, \mathcal{F}_{cs}$ | vectors of magnetomotive forces produced by the stator phase currents, A/ph |
| $\mathcal{F}_s^s$ | vector of stator magnetomotive force in the stator reference frame, A |
| $\mathcal{F}_s$ | magnitude of the vector of stator magnetomotive force, A |
| $\mathcal{F}_{DS}^e, \mathcal{F}_{QS}^e$ | components of the vector of stator magnetomotive force in the excitation reference frame, A |
| $\mathcal{F}_{ds}^s, \mathcal{F}_{qs}^s$ | components of the vector of stator magnetomotive force in the stator reference frame, A |
| $f$ | supply frequency, Hz |
| $f_{rat}$ | rated supply frequency, Hz |
| $h$ | current tolerance band of an inverter, A |
| $\hat{I}_M$ | phasor of magnetizing current in the steady-state $\Gamma$ equivalent circuit, A/ph |
| $\hat{I}_M'$ | phasor of magnetizing current in the steady-state $\Gamma'$ equivalent circuit, A/ph |
| $\hat{I}_m$ | phasor of magnetizing current in the steady-state T equivalent circuit, A/ph |
| $\hat{I}_R$ | phasor of rotor current in the steady-state $\Gamma$ equivalent circuit, A/ph |
| $\hat{I}_R'$ | phasor of rotor current in the steady-state $\Gamma'$ equivalent circuit, A/ph |
| $\hat{I}_r$ | phasor of rotor current in the steady-state T equivalent circuit, A/ph |
| $\hat{I}_s$ | phasor of stator current, A/ph |
| $I_{dc}$ | d.c. supply current of a current source inverter, A |
| $I_M$ | r.m.s. value of magnetizing current in the steady-state $\Gamma$ equivalent circuit, A/ph |
| $I_M'$ | r.m.s. value of magnetizing current in the steady-state $\Gamma'$ equivalent circuit, A/ph |
| $I_m$ | r.m.s. value of magnetizing current in the steady-state T equivalent circuit, A/ph |
| $I_{max}$ | amplitude of fundamental output current of an inverter, A |
| $I_R$ | r.m.s. value of rotor current in the steady-state $\Gamma$ equivalent circuit, A/ph |
| $I_R'$ | r.m.s. value of rotor current in the steady-state $\Gamma'$ equivalent circuit, A/ph |
| $I_r$ | r.m.s. value of rotor current in the steady-state T equivalent circuit, A/ph |
| $I_s$ | r.m.s. value of stator current, A/ph |
| $I_{s,all}$ | maximum allowable r.m.s. value of stator current, A/ph |
| $I_{s,max}$ | peak value of stator current, A/ph |
| $I_{s,rat}$ | rated r.m.s. value of stator current, A/ph |
| $I_{s\Phi}$ | r.m.s. value of flux-producing current, A/ph |
| $I_{sT}$ | r.m.s. value of torque-producing current, A/ph |
| $i_{a1}$ | fundamental phase-A current of an inverter, A |

# NOMENCLATURE

| | |
|---|---|
| $\vec{i}_M^s$ | vector of magnetizing current in the dynamic Γ equivalent circuit and stator reference frame, A |
| $\vec{i}_M^{s\prime}$ | vector of magnetizing current in the dynamic Γ' equivalent circuit and stator reference frame, A |
| $\vec{i}_m^s$ | vector of magnetizing current in the dynamic T equivalent circuit and stator reference frame, A |
| $\vec{i}_R^e$ | vector of rotor current in the excitation reference frame, A |
| $\vec{i}_R^s$ | vector of rotor current in the dynamic Γ equivalent circuit and stator reference frame, A |
| $\vec{i}_R^{s\prime}$ | vector of rotor current in the dynamic Γ' equivalent circuit and stator reference frame, A |
| $\vec{i}_r^r$ | vector of actual rotor current, A |
| $\vec{i}_r^s$ | vector of rotor current in the dynamic T equivalent circuit and stator reference frame, A |
| $\vec{i}_S^e$ | vector of stator current in the excitation frame, A |
| $\vec{i}_s^s$ | vector of stator current in the stator reference frame, A |
| $i_a, i_b, i_c$ | output currents of an inverter, A |
| $i_{a1}$ | fundamental current in phase A of an inverter, A |
| $i_{as}, i_{bs}, i_{cs}$ | stator phase currents, A |
| $i_{DR}^e, i_{QR}^e$ | components of the vector of rotor current in the excitation reference frame, A |
| $i_{DS}^e, i_{QS}^e$ | components of the vector of stator current in the excitation reference frame, A |
| $i_{dr}^s, i_{qr}^s$ | components of the vector of rotor current in the stator reference frame, A |
| $i_{ds}^s, i_{qs}^s$ | components of the vector of stator current in the stator reference frame, A |
| $i_r$ | magnitude of the vector of rotor current, A |
| $i_s$ | magnitude of the vector of stator current, A. |
| $J_M$ | mass moment inertia of the rotor, kg m² |
| $J_L$ | mass moment inertia of the load, kg m² |
| $k_T$ | torque constant, N m/Wb/A |
| $L_L$ | leakage inductance in the Γ equivalent circuit, H/ph |
| $L_L'$ | leakage inductance in the Γ' equivalent circuit, H/ph |
| $L_{lr}$ | rotor leakage inductance, H/ph |
| $L_{ls}$ | stator leakage inductance, H/ph |
| $L_M$ | mutual inductance in the Γ equivalent circuit, H/ph |
| $L_M'$ | mutual inductance in the Γ' equivalent circuit, H/ph |
| $L_m$ | mutual inductance in the T equivalent circuit, H/ph |
| $L_R$ | rotor inductance in the Γ equivalent circuit, H/ph |
| $L_r$ | rotor inductance in the T equivalent circuit, H/ph |
| $L_s$ | stator inductance in the T equivalent circuit, H/ph |
| $L_s'$ | stator inductance in the Γ' equivalent circuit, H/ph |
| $M$ | modulation index (magnitude control ratio) of an inverter |
| $N$ | number of switching intervals per cycle of the output voltage of an inverter |

## NOMENCLATURE

| | |
|---|---|
| $N_p$ | number of pulses of a switching variable per cycle of the output voltage of an inverter |
| $N_r$ | number of turns in the rotor winding (per phase) |
| $N_s$ | number of turns in the stator winding (per phase) |
| $n_M$ | speed, r.p.m. |
| $n_{M,rat}$ | rated speed, r.p.m. |
| $n_{M,syn}$ | synchronous speed, r.p.m. |
| $P$ | number of poles |
| $P_{elec}$ | electrical power, W |
| $P_{inp}$ | input power, W |
| $P_{mech}$ | mechanical power, W |
| $P_{out}$ | output power, W |
| $PF$ | power factor |
| $p$ | differentiation operator ($d/dt$), sec$^{-1}$ |
| $R_R$ | rotor resistance in the $\Gamma$ equivalent circuit, $\Omega$/ph |
| $R_R'$ | rotor resistance in the $\Gamma'$ equivalent circuit, $\Omega$/ph |
| $R_r$ | rotor resistance in the T equivalent circuit, $\Omega$/ph |
| $R_r^r$ | actual rotor resistance, $\Omega$/ph |
| $R_s$ | stator resistance, $\Omega$/ph |
| $s$ | slip |
| $T$ | developed torque, N m |
| $T_L$ | load torque, N m |
| $T_{peak}$ | peak torque (breakdown torque, pull-out torque), N m |
| $T_{rat}$ | rated torque, N m |
| $T_{start}$ | starting torque, N m |
| $t$ | time, sec |
| $\hat{V}_s$ | phasor of stator voltage, V/ph |
| $V_{dc}$ | d.c. supply voltage of a voltage source inverter, V |
| $V_s$ | r.m.s. value of stator voltage, V/ph |
| $V_{s,rat}$ | rated r.m.s. value of stator voltage, V/ph |
| $V_{max}$ | maximum available magnitude of the vector of output voltage of an inverter, V |
| $v$ | vector of output voltage of an inverter, V |
| $v_X, v_Y, v_Z$ | vectors of output voltage of an inverter corresponding to states $X, Y, Z$, V |
| $v_r^r$ | vector of actual rotor voltage, V |
| $v_r^s$ | vector of rotor voltage in the stator reference frame, V |
| $v_s^e$ | vector of stator voltage in the excitation reference frame, V |
| $v_s^s$ | vector of stator voltage in the stator reference frame, V |
| $v_a, v_b, v_c$ | line-to-neutral output voltages of an inverter, V |
| $v_{ab}, v_{bc}, v_{ca}$ | line-to-line output voltages of an inverter, V |
| $v_{as}, v_{bs}, v_{cs}$ | stator phase voltages, V |

# NOMENCLATURE

| | |
|---|---|
| $v_{DS}^e, v_{QS}^e$ | components of the vector of stator voltage in the excitation reference frame, V |
| $v_{ds}^s, v_{qs}^s$ | components of the vector of rotor voltage in the stator reference frame, V |
| $v_{ds}^s, v_{qs}^s$ | components of the vector of stator voltage in the stator reference frame, V |
| $X_{ls}$ | rotor leakage reactance, $\Omega$/ph |
| $X_{ls}$ | stator leakage reactance, $\Omega$/ph |
| $X_m$ | magnetizing reactance, $\Omega$/ph |
| $X_r$ | rotor reactance, $\Omega$/ph |
| $X_s$ | stator reactance, $\Omega$/ph |
| $\gamma$ | coefficient of transformation of the T equivalent circuit to the $\Gamma$ equivalent circuit |
| $\gamma'$ | coefficient of transformation of the T equivalent circuit to the $\Gamma'$ equivalent circuit |
| $\eta$ | efficiency |
| $\Theta$ | time integral of slip speed, rad |
| $\Theta_M$ | angular displacement of the rotor of a $P$-pole motor, rad |
| $\Theta_m$ | angular position (phase) of the vector of airgap flux in the stator reference frame, rad |
| $\Theta_o$ | angular displacement of the rotor of a 2-pole motor, rad |
| $\Theta_r$ | angular position (phase) of the vector of rotor flux in the stator reference frame, rad |
| $\Theta_s$ | angular position (phase) of the vector of stator flux in the stator reference frame, rad. Also, that of the vector of stator magnetomotive force, rad. |
| $\hat{\Lambda}_R$ | phasor of rotor flux in the steady-state $\Gamma$ equivalent circuit, Wb/ph |
| $\hat{\Lambda}_R'$ | phasor of rotor flux in the steady-state $\Gamma'$ equivalent circuit, Wb/ph |
| $\hat{\Lambda}_r$ | phasor of rotor flux in the steady-state T equivalent circuit, Wb/ph |
| $\hat{\Lambda}_s$ | phasor of stator flux, Wb/ph |
| $\Lambda_R$ | r.m.s. value of rotor flux in the steady-state $\Gamma$ equivalent circuit, Wb/ph |
| $\Lambda_R'$ | r.m.s. value of rotor flux in the steady-state $\Gamma'$ equivalent circuit, Wb/ph |
| $\Lambda_s$ | r.m.s. value of stator flux, Wb/ph |
| $\lambda_M^e$ | vector of airgap flux in the excitation reference frame, Wb |
| $\lambda_m^s$ | vector of airgap flux in the dynamic T equivalent circuit and stator reference frame, Wb |
| $\lambda_R^e$ | vector of rotor flux in the excitation reference frame, Wb |
| $\lambda_R^s$ | vector of rotor flux in the dynamic $\Gamma$ equivalent circuit and stator reference frame, Wb/ph |
| $\lambda_R^{s'}$ | vector of rotor flux in the dynamic $\Gamma'$ equivalent circuit and stator reference frame, Wb/ph |
| $\lambda_r^r$ | vector of actual rotor flux, Wb |
| $\lambda_r^s$ | vector of rotor flux in the dynamic T equivalent circuit and stator reference frame, Wb |
| $\lambda_S^e$ | vector of stator flux in the excitation reference frame, Wb |
| $\lambda_s^s$ | vector of stator flux in the stator reference frame, Wb |

# NOMENCLATURE

| | |
|---|---|
| $\lambda_{DM}^e, \lambda_{QM}^e$ | components of the vector of airgap flux in the excitation reference frame, Wb |
| $\lambda_{DR}^e, \lambda_{QR}^e$ | components of the vector of rotor flux in the excitation reference frame, Wb |
| $\lambda_{DS}^e, \lambda_{QS}^e$ | components of the vector of stator flux in the excitation reference frame, Wb |
| $\lambda_{dm}^s, \lambda_{qm}^s$ | components of the vector of airgap flux in the stator reference frame, Wb |
| $\lambda_{dr}^s, \lambda_{qr}^s$ | components of the vector of rotor flux in the stator reference frame, Wb |
| $\lambda_{ds}^s, \lambda_{qs}^s$ | components of the vector of stator flux in the stator reference frame, Wb |
| $\lambda_m$ | magnitude of the vector of airgap flux, Wb |
| $\lambda_r$ | magnitude of the vector of rotor flux, Wb |
| $\lambda_s$ | magnitude of the vector of stator flux, Wb |
| $\sigma$ | total leakage factor |
| $\sigma_r$ | rotor leakage factor |
| $\nu$ | turns ratio |
| $\tau$ | ratio of leakage inductance to rotor resistance in the $\Gamma$ equivalent circuit |
| $\tau_r$ | rotor time constant, i.e., ratio of rotor inductance to rotor resistance in the T equivalent circuit |
| $\omega$ | supply radian frequency, synchronous speed of a 2-pole motor, rad/sec |
| $\omega_{rat}$ | rated supply radian frequency, rad/sec |
| $\omega_M$ | speed of a $P$-pole motor, rad/sec |
| $\omega_o$ | speed of a 2-pole motor, rad/sec |
| $\omega_r$ | slip speed of a 2-pole motor, rad/sec |
| $\omega_{sl}$ | slip speed of a $P$-pole motor, rad/sec |
| $\omega_{syn}$ | synchronous speed of a $P$-pole motor, rad/sec |

## Subscripts

| | |
|---|---|
| a, b, c | phase A, phase B, phase C |
| *all* | maximum allowable |
| D | $D$-axis (excitation reference frame) |
| d | $d$-axis (stator reference frame) |
| dc | d.c. |
| L, l | leakage |
| *L* | load |
| M, m | mutual, magnetizing |
| *M* | motor |
| *max* | amplitude |
| o | rotor (2-pole motor) |
| Q | $Q$-axis (excitation reference frame) |
| q | $q$-axis (stator reference frame) |
| R, r | rotor (with the exception of the slip speed, $\omega_r$) |

NOMENCLATURE

| | |
|---|---|
| *rat* | rated |
| S, s | stator |
| *sl* | slip |
| *syn* | synchronous |
| T, $T$ | torque |
| X, Y, Z | state $X$, state $Y$, state $Z$ of an inverter |

## Superscripts

| | |
|---|---|
| **e**, e | excitation reference frame |
| **r**, r | actual rotor quantities |
| **s**, s | stator reference frame |
| * | conjugate or reference |
| ' | $\Gamma$ equivalent circuit (unless otherwise specified) |

## Abbreviations

| | |
|---|---|
| BJT | Bipolar Junction Transistor |
| CSI | Current Source Inverter |
| CVH | Constant Volts/Hertz |
| FOP | Field Orientation Principle |
| GTO | Gate Turn-Off Thyristor |
| IGBT | Insulated Gate Bipolar Transistor |
| MCT | MOS-Controlled Thyristor |
| MOSFET | Metal-Oxide Semiconductor Field-Effect Transistor |
| MRAS | Model Reference Adaptive System |
| PI | Proportional-Plus-Integral (controller) |
| PID | Proportional-Plus-Integral-Plus-Derivative (controller) |
| PWM | Pulse Width Modulation |
| RPWM | Random Pulse Width Modulation |
| SCR | Silicon Controlled Rectifier |
| SIT | Static Induction Transistor |
| SITH | Static Induction Thyristor |
| SOAR | Safe Operating Area |
| TC | Torque Control |
| UFO | Universal Field Orientation |
| VSC | Variable Structure Control |
| VSI | Voltage Source Inverter |

# Preface

Induction motors, particularly those of the squirrel-cage type, have been for almost a century industry's principal workhorse. Until the early seventies, they had been mostly operated in the constant-frequency, constant-voltage, uncontrolled mode which even today is still most common in practice. Adjustable-speed drives had been based on d.c. motors, mainly in the classic Ward-Leonard arrangement.

The advent of thyristors, the first controlled semiconductor power switches and, consequently, the development of variable-frequency converters based on these switches, made possible wide-range speed control of induction motors. The most popular, so-called scalar control methods consist in simultaneous adjustments of the frequency and magnitude of the sinusoidal voltage or current supplied to the motor. This allows making steady-state operating characteristics of an induction motor similar to those of a d.c. motor. Adjustable-speed a.c. drive systems, employing scalar control principles have been replacing the d.c. drives in numerous industrial applications, such as pumps, fans, compressors, and conveyor belts. Induction motors have here a clear competitive edge over d.c. machines. They are significantly less expensive, more robust, and capable of reliable operation in harsh ambient conditions, even in an explosive atmosphere.

It must be stressed, that scalar control does not make an induction motor dynamically equivalent to a d.c. motor. Accurate position control, for instance, is not possible, since precise control of the instantaneous torque developed in the motor is needed to realize the required speed trajectory. This, in turn, means that it is the instantaneous stator current that has to be controlled. Scalar control applies to steady-state operation of the motor only, while no special allowance for transient operating conditions appears in the control principles.

Brushes, affixed to the stator, and commutator on the rotor provide in a d.c. motor a direct, physical link between the field and armature circuits of the machine. Proper positioning of the brushes ensures optimal conditions for torque production under all operating conditions. This is not the case in an induction motor, whose rotor is physically isolated from the stator. Therefore, to optimize the torque production conditions in this motor, an algorithmic equivalent of the d.c. motor's physical stator-rotor link has to be provided by an appropriate control technique. Such techniques are based on the, so-called, Field Orientation Principle which is the topic of this book.

# PREFACE

The Field Orientation Principle was first formulated by Haase, in 1968, and Blaschke, in 1970. At that time, their ideas seemed impractical because of the insufficient means of implementation. However, in the early eighties, technological advances in static power converters and microprocessor-based control systems made the high-performance a.c. drive systems fully feasible. Since then, hundreds of papers dealing with various aspects of the Field Orientation Principle have appeared every year in the technical literature, and numerous commercial high-performance a.c. drives based on this principle have been developed. The term "vector control" is often used with regard to these systems. Today, it seems certain that almost all d.c. industrial drives will be ousted in the foreseeable future, to be, in major part, superseded by a.c. drive systems with vector controlled induction motors. This transition has already been taking place in industries of developed countries. Vector controlled a.c. drives have been proven capable of even better dynamic performance than d.c. drive systems, because of higher allowable speeds and shorter time constants of a.c. motors.

It should be mentioned that the Field Orientation Principle can be used in control not only of induction (asynchronous) motors, but of all kinds of synchronous motors as well. Vector controlled drive systems with the so-called brushless d.c. motors have found many applications in high-performance drive systems, such as machine tools and industrial robots. Large, cycloconverter-fed synchronous motors can also be controlled on the basis of the Field Orientation Principle. The same applies to recently developed synchronous reluctance motors. This book, however, is focused on induction motors only, on the assumption that these machines represent a distinct majority in industrial drive systems. Actually, field orientation in synchronous motors is simpler than in induction motors, since the position of the flux vector generated in the rotor is easy to monitor. Therefore, interested readers should not encounter any serious difficulties in extending the acquired knowledge of the Field Orientation Principle to the other kinds of a.c. motors. It is also worth mentioning that there exist concepts of vector control which do not directly employ the Field Orientation Principle, such as the theory of spiral vectors and field acceleration. It is too early to judge whether these alternative approaches to control of a.c. motors will gain popularity in the commercial practice.

It is assumed that the reader is familiar with the basic concepts of electromechanical energy conversion and electric machines. The Field Orientation Principle is presented both in a formal manner, based on the vector theory of a.c. machines, and in simple engineering terms. In Chapter 1, dynamic models, based on the concept of space vectors of motor quantities, are compared with the well known phasor-based, steady-state equivalent circuits of an induction machine. Explanation of the relation between vectors and phasors allows viewing the per-phase, steady-state equivalent circuit of the motor as a special case of the dynamic model. Electromagnetic matrix equations and the torque equation, in both the stator (stationary) and excitation (rotating) reference frames, are then derived. They provide means for computer simulation of induction motors and serve as a background for the Field Orientation Principle.

Before the introduction of the field orientation concept, scalar control techniques for adjustable-speed, induction-motor drives are described in Chapter 2. Next, in Chapter 3, control characteristics of a separately excited d.c. motor are explained, and fundamentals of the Field Orientation Principle are introduced. Chapter 4 presents the Field Orientation Principle as a means of optimizing the torque-production conditions and of decoupling the flux and torque control in an induction motor. Both the direct and indirect approaches to rotor flux orientation in induction motors are described. Chapter 5 is devoted to power inverters, which constitute the controllable supply source for induction motors. Chapter 6, which concludes the book, contains a review of vector control systems for induction motors. Stator and airgap flux orientation schemes, systems with current source inverters, observers of motor variables, estimation of motor parameter, adaptive tuning, and speed and position control systems for field-oriented induction motors are briefly examined there.

To facilitate the reader's understanding of the material presented, numerous numerical examples and computer simulations are included. It must be stressed that the book is not intended to exhaustively cover the latest achievements in the area of high-performance a.c. drives, and it is not an engineering manual. It is primarily addressed to practicing engineers, graduate students in power and control programs, and those researchers who intend to enter the field of induction motor control but lack a sufficient background in the dynamics of these motors. Hence, the book should be viewed as a graduate-level textbook rather than as a sophisticated scientific monograph.

The book has originated from the lecture notes prepared by the author for a graduate course in electric drive systems and power electronics, taught from 1990 at the University of Nevada, Reno. The existing books and papers have been found either too superficial or too complicated for readers familiar with only the steady-state theory of electrical machines, to which typical undergraduate curricula are limited. It is hoped that this book will fill that gap.

PREFACE

The author wants to acknowledge the invaluable help of Dr. Ronald E. Colyer of the Royal Military College of Science in Great Britain, whose precious comments and suggestions greatly enhanced quality of the book. Expressions of gratitude and apology are directed to the author's wife, Dorota, and children, Bart and Nicole, who patiently endured the long working hours of their husband and father.

# The Field Orientation Principle in Control of Induction Motors

# Chapter 1

# DYNAMIC MODEL OF THE INDUCTION MOTOR

The well known equivalent circuit of the induction motor allows calculation of the basic quantities of a given motor, such as stator current, power factor, developed torque, etc., when the motor operates in the steady state, with constant speed and fixed, balanced, sinusoidal supply voltage. The electrical quantities are represented as phasors and the developed torque is calculated as the output power divided by the angular velocity of the rotor. Since the output power is computed as an average (per cycle), and not as an instantaneous, quantity, it follows that it is also an average, not an instantaneous, value of the torque that is determined. The equivalent circuit is, therefore, insufficient for analysis of transient operating conditions. Also, if a non-sinusoidal voltage, typical for power electronic converters, is applied to the stator then, even in the steady state, calculation of the motor performance using the harmonic decomposition and superposition principles is very tedious.

In this chapter, two versions of the dynamic model of an induction motor are developed, one in the stator reference frame and one in the excitation frame. The dynamic model is based on the concept of vector quantities of an a.c. machine, introduced by Kovacs and Racz in 1959. The motor can be represented either in the form of an equivalent circuit or a set of equations. This procedure allows analysis of the dynamics of the motor which can then be supplied with any kind of voltage, not necessarily a sinusoidal one. Although the analysis is generally performed by means of digital simulation, certain important features of the motor dynamics can be deduced directly from the model.

## 1.1 Space Vectors in Stator Reference Frame

Figure 1.1 shows, schematically, a cross-section of the stator of a three-phase, two-pole a.c. motor. For simplicity, it is assumed that each phase winding consists of one single-turn coil. A total of six conductors (coil sides) are thus embedded in the iron of the stator. Practical motors have more complex stator windings, hence the simple model in question will further be called a primitive three-phase motor. A phase current is assumed positive if it enters the conductor designated by an unprimed phase denomination, e.g., $A$, and it is assumed negative if it enters the conductor designated by a primed phase denomination, e.g., $A'$. Two axes, the direct axis, $d$, and the quadrature axis, $q$, are aligned with the horizontal and vertical geometrical axes of the stator, respectively. They represent the so-called stator, or stationary, reference frame.

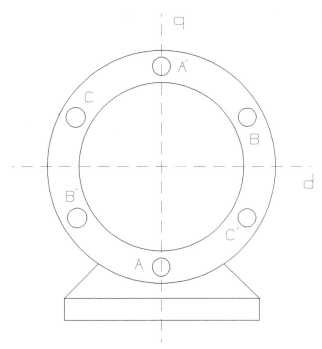

**Figure 1.1.** Schematic representation of a three-phase, 2-pole stator.

Initially, it is assumed that the stator coils are supplied from a balanced three-phase a.c. source with a radian frequency $\omega$. In Figure 1.2, the stator is shown together with a phasor diagram of the stator currents, $\hat{I}_{as}$, $\hat{I}_{bs}$ and $\hat{I}_{cs}$, at an initial instant of time, i.e., at $\omega t = 0°$. Polarities of currents in the coils, derived from the phasor diagram, are marked using the standard system of dots and crosses.

Currents $i_{as}$, $i_{bs}$, and $i_{cs}$ in the stator coils $AA'$, $BB'$ and $CC'$ produce magnetomotive forces (mmf's) $\mathscr{F}_{as}$, $\mathscr{F}_{bs}$ and $\mathscr{F}_{cs}$ which are space vectors, as shown in Figure 1.2. The resultant stator mmf vector, $\mathscr{F}_{s}$, constitutes a vector sum of the phase mmf vectors. Here, and throughout the book, a bold font denotes space vectors, while a superscript indicates the reference frame.

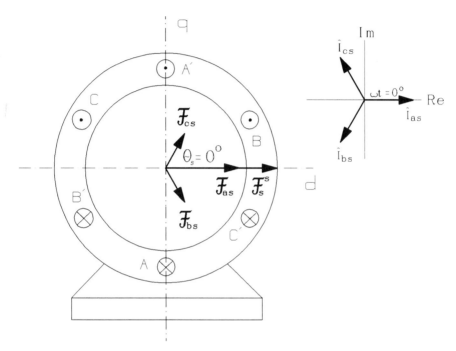

**Figure 1.2.** Stator mmf vectors at $\omega t = 0°$.

Figure 1.3 illustrates a situation when the current phasors have rotated by 60° with respect to those shown in Figure 1.2, i.e., one sixth of a cycle later. It can be seen that vector $\boldsymbol{\mathcal{F}}_s^s$ has turned in the stator reference frame by the same angle of 60°. Its magnitude has remained unchanged, at 1.5 times the magnitude of the phase mmfs. It must be stressed that the vector quantities appear in the real, physical space of the motor, while the phasors constitute an abstract representation of physical a.c. quantities in a fictitious, complex plane. Therefore, the geometrical angle, $\Theta_s$, equals the electrical angle, $\omega t$, only in a 2-pole stator. Generally, in a $P$-pole stator,

$$\Theta_s = \frac{2}{P}\omega t. \qquad (1.1)$$

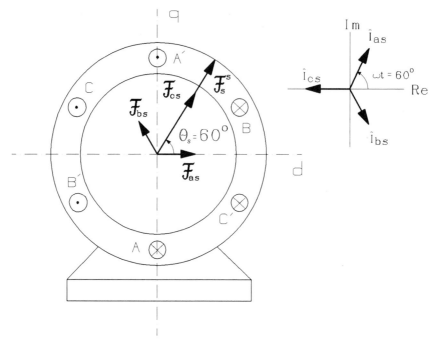

**Figure 1.3.** Stator mmf vectors at $\omega t = 60°$.

Taking the $d$-axis as real and the $q$-axis as imaginary, vector $\mathcal{F}_s^s$ can be expressed as

$$\mathcal{F}_s^s = \mathcal{F}_s e^{j\theta_s} = \mathcal{F}_{ds}^s + j\mathcal{F}_{qs}^s \qquad (1.2)$$

where, as illustrated in Figure 1.4, $\mathcal{F}_{ds}^s$ and $\mathcal{F}_{qs}^s$ are its horizontal ($d$-axis) and vertical ($q$-axis) components, while $\mathcal{F}_s$ denotes the magnitude of vector $\mathcal{F}_s^s$. It has to be stressed that the magnitude of a vector does not depend on the reference frame.

In the steady-state operating mode of the primitive motor considered so far, vector $\mathcal{F}_s^s$ has a constant magnitude and rotates with angular velocity $\omega$, equal to the supply radian frequency. Stator currents which are not balanced or sinusoidal, will produce a certain mmf vector $\mathcal{F}_s^s$ with a magnitude and/or speed of rotation which are not necessarily constant. Therefore, the vector representation is more general than the phasor representation which applies only to sinusoidal, constant-magnitude and constant-frequency quantities.

# DYNAMIC MODEL OF THE INDUCTION MOTOR

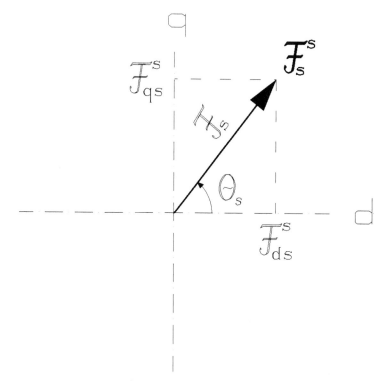

**Figure 1.4.** Stator mmf components in the stator reference frame.

The rotating mmf vector $\mathscr{F}_s^s$ in Figures 1.2 and 1.3 was determined by vector addition of stationary, although pulsating, vectors $\mathscr{F}_{as}$, $\mathscr{F}_{bs}$, and $\mathscr{F}_{cs}$. Since the latter vectors are perpendicular to their correponding phase coils which are 120° apart from each other, and since the phase-A coil was assumed to be in the vertical axis of the stator, then, in analytical form,

$$\mathscr{F}_s^s = \mathscr{F}_{as}e^{j0°} + \mathscr{F}_{bs}e^{j120°} + \mathscr{F}_{cs}e^{j240°}. \qquad (1.3)$$

The concept of space vectors can be extended to other quantities of the motor, such as currents, voltages, and flux linkages. In fact, in the primitive motor, due to the single-turn coils, the stator current vector, $\vec{i}_s^s$, is identical to the stator mmf vector, $\mathscr{F}_s^s$, and the components, $i_{ds}^s$ and $i_{qs}^s$, of the current vector have a real, practical meaning. Specifically, they represent the stator currents of a two-phase primitive a.c. motor, equivalent to the three-phase model under consideration. This observation is illustrated in Figures 1.5 and 1.6 which correspond to Figures 1.2 and 1.3, respectively.

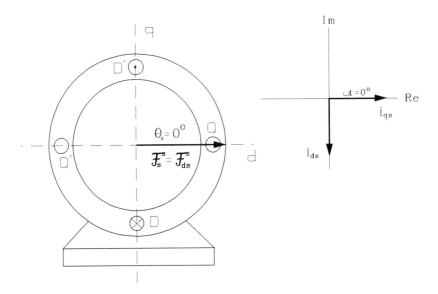

**Figure. 1.5.** Mmf vectors of a two-phase stator at $\omega t = 0°$.

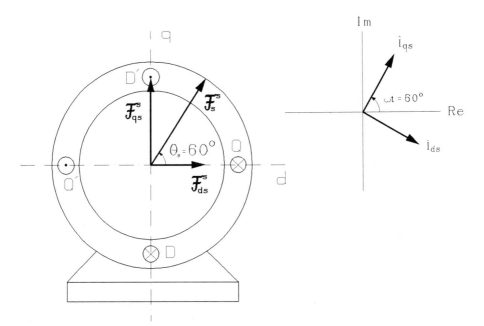

**Figure 1.6.** Mmf vectors of a two-phase stator at $\omega t = 60°$.

# DYNAMIC MODEL OF THE INDUCTION MOTOR

The other stator vectors to be used in further considerations are the voltage vector

$$\mathbf{v_s^s} = v_{as}^s e^{j0°} + v_{bs}^s e^{j120°} + v_{cs}^s e^{j240°} \qquad (1.4)$$

and the flux (actually, flux-linkage) vector

$$\mathbf{\lambda_s^s} = \lambda_{as}^s e^{j0°} + \lambda_{bs}^s e^{j120°} + \lambda_{cs}^s e^{j240°}. \qquad (1.5)$$

Voltages $v_{as}$, $v_{bs}$, and $v_{cs}$ constitute the supply voltages of the individual phase windings of the stator. They are taken as line-to-neutral voltages if the windings are connected in wye (star) and line-to-line voltages for a delta (mesh) connection. Variables $\lambda_{as}$, $\lambda_{bs}$, and $\lambda_{cs}$ represent magnetic fluxes linking these windings.

The problem of rotor quantities is somewhat complicated by the fact that the rotor itself revolves in the stator reference frame. Hence, the rotor vectors rotate both round the rotor and with the rotor. In the steady state, for instance, the rotor currents of radian frequency $\omega_r$ produce in the primitive, 2-pole, motor the mmf vector revolving round the rotor with the same angular speed $\omega_r$. Denoting the angular speed of the rotor by $\omega_o$, it can be seen that with respect to the stator reference frame the rotor mmf rotates with the combined angular speed

$$\omega_r + \omega_o = \omega \qquad (1.6)$$

i.e., the same speed as that of the stator mmf. Variable $\omega_r$ will further be referred to as slip speed as representing, in a 2-pole motor, the difference between the speed of the revolving magnetic field of the stator (synchronous speed), $\omega$, and the speed of the rotor (motor speed), $\omega_o$.

It must be pointed out that the variously subscripted symbols "$\omega$" used throughout the book denote both the radian frequency, in the electrical sense, and the angular velocity (speed), in the mechanical sense. For example, currents of frequency 60 Hz, i.e., radian frequency of 377 rad/sec, result in a 2-pole stator in the mmf vector rotating with the same angular speed of 377 rad/sec.

In a real motor, the rotor winding differs from the stator winding, i.e., the effective number of turns per phase of the rotor winding, $N_r$, is not equal to that of the stator winding $N_s$. Therefore, in order to express the rotor current, voltage, and flux vectors in the stator reference frame, both the rotor angular position, $\Theta_o$, with respect to the $d$-axis and the turns ratio, $v = N_s/N_r$, must be taken into account.

The transformation of actual rotor vectors $\vec{i}_r^r$, $\vec{v}_r^r$, and $\vec{\lambda}_r^r$ into vectors $\vec{i}_r^s$, $\vec{v}_r^s$, and $\vec{\lambda}_r^s$ in the stator reference frame is, in respect of the turns ratio, the same as that used in the steady-state theory of transformers and induction motors. In such transformations, the actual rotor current phasor is divided by $v$, while the actual voltage and flux phasors are multiplied by $v$. When the rotational operator $exp(j\Theta_o)$ is additionally employed to account for the rotor motion, the following equations result:

$$\vec{i}_r^s = \frac{e^{j\Theta_o}}{v} \vec{i}_r^r \qquad (1.7)$$

$$\vec{v}_r^s = v e^{j\Theta_o} \vec{v}_r^r \qquad (1.8)$$

$$\vec{\lambda}_r^s = v e^{j\Theta_o} \vec{\lambda}_r^r. \qquad (1.9)$$

It is worth mentioning that the rotor-to-stator vector transformations described above, necessary in the course of development of the dynamic model of induction motor, are, otherwise, of little practical value. The actual rotor quantities appearing in Eqs. (1.7) through (1.9) are usually unknown because of the lack of access to the rotor in the squirrel-cage induction motors, typically used in drive systems based on the Field Orientation Principle. For all practical purposes, rotor quantities referred to the stator are employed, and their relations to the real rotor quantities are not relevant. Hence, a knowledge of, for instance, the turns ratio is not necessary for the design of a control system of a given motor.

## 1.2 Direct and Inverse Three-Phase to Stator Reference Frame Transformations

Replacing the stator mmf's in Eq. (1.6) with stator phase currents $i_{as}$, $i_{bs}$, and $i_{cs}$, and employing Euler's identity $e^{j\omega t} = cos(\omega t) + jsin(\omega t)$, the abstract $i_{ds}^s$ and $i_{qs}^s$ components of the vector $\vec{i}_s^s$ can be expressed in terms of the instantaneous values of the actual phase currents as

# DYNAMIC MODEL OF THE INDUCTION MOTOR

the instantaneous values of the actual phase currents as

$$\begin{aligned}\mathbf{i}_s^s &= i_{as}[\cos(0°)+j\sin(0°)] \\ &\quad + i_{bs}[\cos(120°)+j\sin(120°)] \\ &\quad + i_{cs}[\cos(240°)+j\sin(240°)] \\ &= i_{as} - \frac{1}{2}i_{bs} - \frac{1}{2}i_{cs} \\ &\quad + j(\frac{\sqrt{3}}{2}i_{bs} - \frac{\sqrt{3}}{2}i_{cs}) \\ &= i_{ds}^s + j i_{qs}^s .\end{aligned} \quad (1.10)$$

Hence,

$$\begin{bmatrix} i_{ds}^s \\ i_{qs}^s \end{bmatrix} = \begin{bmatrix} 1 & -\frac{1}{2} & -\frac{1}{2} \\ 0 & \frac{\sqrt{3}}{2} & -\frac{\sqrt{3}}{2} \end{bmatrix} \begin{bmatrix} i_{as} \\ i_{bs} \\ i_{cs} \end{bmatrix}. \quad (1.11)$$

The derived matrix equation expresses the so-called *abc→dq* transformation, i.e., the transformation of phase quantities, in this case phase currents, into corresponding vectors in the stator reference frame. The matrix in Eq. (1.11) is not square, therefore an inverse transformation requires the addition of the third row. Assuming a three-wire supply to the motor, the phase currents add up to a zero, i.e.,

$$i_{as} + i_{bs} + i_{cs} = 0 \quad (1.12)$$

which leads to the following expansion of the matrix equation (1.11):

$$\begin{bmatrix} i_{ds}^s \\ i_{qs}^s \\ 0 \end{bmatrix} = \begin{bmatrix} 1 & -\frac{1}{2} & -\frac{1}{2} \\ 0 & \frac{\sqrt{3}}{2} & -\frac{\sqrt{3}}{2} \\ 1 & 1 & 1 \end{bmatrix} \begin{bmatrix} i_{as} \\ i_{bs} \\ i_{cs} \end{bmatrix}. \quad (1.13)$$

Now, the inverse, $dq \rightarrow abc$ transformation can easily be performed. It is expressed by the matrix equation

$$\begin{bmatrix} i_{as} \\ i_{bs} \\ i_{cs} \end{bmatrix} = \begin{bmatrix} \frac{2}{3} & 0 & \frac{1}{3} \\ -\frac{1}{3} & \frac{1}{\sqrt{3}} & \frac{1}{3} \\ -\frac{1}{3} & -\frac{1}{\sqrt{3}} & \frac{1}{3} \end{bmatrix} \begin{bmatrix} i_{ds}^{s} \\ i_{qs}^{s} \\ 0 \end{bmatrix} \quad (1.14)$$

or

$$\begin{bmatrix} i_{as} \\ i_{bs} \\ i_{cs} \end{bmatrix} = \begin{bmatrix} \frac{2}{3} & 0 \\ -\frac{1}{3} & \frac{1}{\sqrt{3}} \\ -\frac{1}{3} & -\frac{1}{\sqrt{3}} \end{bmatrix} \begin{bmatrix} i_{ds}^{s} \\ i_{qs}^{s} \end{bmatrix}. \quad (1.15)$$

Analogous transformations apply to the other vector quantities of the motor. It should be pointed out that the third row of the last matrix in Eq. (1.14), here assumed zero, represents the so-called zero-sequence component, used in analysis of unbalanced three-phase circuits. If the stator windings were supplied from a four-wire power line, this component would represent the imbalance current in the neutral wire. Such connection is, however, not typical for practical drive systems.

One of the advantages of the use of the space vector form of motor quantities is the demonstrated reduction of three, one per phase, physical variables ($abc$) to two abstract variables ($dq$), which simplifies the analysis of the motor dynamics. This reduction is possible because of constraint (1.12), which is also applicable to the other phase quantities, like voltages or fluxes. The presence of this constraint means that only two out of three phase variables are independent, thereby reducing the number of degrees of freedom for each quantity from three to two.

## 1.3 Voltage and Current Equations in Stator Reference Frame

Using vector notation, either the stator or rotor winding can be represented by a simple resistive-plus-inductive circuit, using current, voltage, and flux space vectors in place of the normally used varying time functions of these quantities. This is illustrated in Figure 1.7, where Faraday's law is used to determine an emf vector $e$ induced in an inductance by a time-varying flux vector $\lambda$.

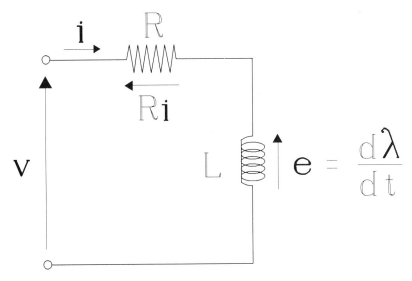

**Figure 1.7.** Resistive-plus-inductive equivalent circuit of either the stator or rotor winding.

Employing the vector version of Kirchhoff's voltage law, the equation of the stator winding can be written as

$$\mathbf{v}_s^s = R_s \mathbf{i}_s^s + \frac{d\boldsymbol{\lambda}_s^s}{dt} \qquad (1.16)$$

and that of the rotor winding as

$$\mathbf{v}_r^r = R_r^r \mathbf{i}_r^r + \frac{d\boldsymbol{\lambda}_r^r}{dt} \qquad (1.17)$$

where $R_s$ and $R_r^r$ are the actual stator and rotor resistances per phase, respectively. As is known from the steady-state theory of induction machines, the relation between $R_r^r$ and the rotor resistance $R_r$ referred to the stator is

$$R_r^r = \frac{1}{v^2} R_r. \qquad (1.18)$$

Hence, from Eqs.(1.7) and (1.18), the first term in Eq. (1.17) is

$$R_r^r \mathbf{1}_r^r = \frac{e^{j\Theta_o}}{v} R_r \mathbf{1}_r^s. \qquad (1.19)$$

The second term, from Eq. (1.9), gives

$$\begin{aligned} \frac{d\boldsymbol{\lambda}_r^r}{dt} &= \frac{d}{dt}(\frac{e^{-j\Theta_o}}{v}\boldsymbol{\lambda}_r^s) \\ &= \frac{1}{v}(\frac{de^{-j\Theta_o}}{dt}\boldsymbol{\lambda}_r^s + \frac{d\boldsymbol{\lambda}_r^s}{dt}e^{-j\Theta_o}) \\ &= \frac{1}{v}[e^{-j\Theta_o}\frac{d(-j\Theta_o)}{dt}\boldsymbol{\lambda}_r^s + \frac{d\boldsymbol{\lambda}_r^s}{dt}e^{-j\Theta_o}] \end{aligned} \qquad (1.20)$$

and, since

$$\frac{d\Theta_o}{dt} = \omega_o \qquad (1.21)$$

then

$$\frac{d\boldsymbol{\lambda}_r^r}{dt} = \frac{e^{-j\Theta_o}}{v}(\frac{d\boldsymbol{\lambda}_r^s}{dt} - j\omega_o \boldsymbol{\lambda}_r^s). \qquad (1.22)$$

Finally, substituting Eqs. (1.19) and (1.22) in Eq. (1.17),

$$\boldsymbol{v}_r^s = R_r \mathbf{1}_r^s + \frac{d\boldsymbol{\lambda}_r^s}{dt} - j\omega_o \boldsymbol{\lambda}_r^s. \qquad (1.23)$$

$$\begin{aligned}
\boldsymbol{v_s^s} &= (R_s+pL_s)(i_{ds}^s+ji_{qs}^s) \\
&\quad +pL_m(i_{dr}^s+ji_{qr}^s) \\
&= (R_s+pL_s)i_{ds}^s+pL_m i_{dr}^s \\
&\quad +j[(R_s+pL_s)i_{qs}^s+pL_m i_{qr}^s].
\end{aligned} \quad (1.28)$$

Clearly, the real and imaginary terms on the right side of the equation represent the *dq* components of vector $v_s^s$, i.e.,

$$v_{ds}^s = (R_s+pL_s)i_{ds}^s+pL_m i_{dr}^s \quad (1.29)$$

$$v_{qs}^s = (R_s+pL_s)i_{qs}^s+pL_m i_{qr}^s. \quad (1.30)$$

Analogously,

$$\begin{aligned}
v_{dr}^s &= pL_m i_{ds}^s+\omega_o L_m i_{qs}^s+(R_r+pL_r)i_{dr}^s \\
&\quad +\omega_o L_r i_{qr}^s
\end{aligned} \quad (1.31)$$

$$\begin{aligned}
v_{qr}^s &= -\omega_o L_m i_{ds}^s+pL_m i_{qs}^s-\omega_o L_r i_{dr}^s \\
&\quad +(R_r+pL_r)i_{qr}^s.
\end{aligned} \quad (1.32)$$

The voltage equation of the motor can now be written as

$$\begin{bmatrix} v_{ds}^s \\ v_{qs}^s \\ v_{dr}^s \\ v_{qr}^s \end{bmatrix} = \begin{bmatrix} R_s+pL_s & 0 & pL_m & 0 \\ 0 & R_s+pL_s & 0 & pL_m \\ pL_m & \omega_o L_m & R_r+pL_r & \omega_o L_r \\ -\omega_o L_m & pL_m & -\omega_o L_r & R_r+pL_r \end{bmatrix} \begin{bmatrix} i_{ds}^s \\ i_{qs}^s \\ i_{dr}^s \\ i_{qr}^s \end{bmatrix}$$

$$(1.33)$$

or

# DYNAMIC MODEL OF THE INDUCTION MOTOR

Introducing a differentiation operator $p \equiv d/dt$, the voltage equations of an induction motor can be written as

$$\mathbf{v}_s^s = R_s \mathbf{i}_s^s + p\boldsymbol{\lambda}_s^s \qquad (1.24)$$

$$\mathbf{v}_r^s = R_r \mathbf{i}_r^s + (p - j\omega_o)\boldsymbol{\lambda}_r^s. \qquad (1.25)$$

The flux vectors $\boldsymbol{\lambda}_s^s$ and $\boldsymbol{\lambda}_r^s$ can then be expressed in terms of current vectors $\mathbf{i}_s^s$ and $\mathbf{i}_r^s$ and the motor inductances as

$$\begin{bmatrix} \boldsymbol{\lambda}_s^s \\ \boldsymbol{\lambda}_r^s \end{bmatrix} = \begin{bmatrix} L_s & L_m \\ L_m & L_r \end{bmatrix} \begin{bmatrix} \mathbf{i}_s^s \\ \mathbf{i}_r^s \end{bmatrix} \qquad (1.26)$$

where

$L_m$    mutual inductance,
$L_s$    stator inductance, calculated as a sum of the stator leakage inductance $L_{ls}$ and mutual inductance $L_m$,
$L_r$    rotor inductance, calculated as a sum of rotor leakage inductance $L_{lr}$ and mutual inductance $L_m$.

Incorporating Eq. (1.26) into Eqs. (1.24) and (1.25), a matrix voltage equation of the motor is obtained as

$$\begin{bmatrix} \mathbf{v}_s^s \\ \mathbf{v}_r^s \end{bmatrix} = \begin{bmatrix} R_s + pL_s & pL_m \\ (p - j\omega_o)L_m & R_r + (p - j\omega_o)L_r \end{bmatrix} \begin{bmatrix} \mathbf{i}_s^s \\ \mathbf{i}_r^s \end{bmatrix}. \qquad (1.27)$$

In the presented form, Eq. (1.27) is unsuitable for dynamic simulation of an induction motor on a digital computer. Therefore, further rearrangement is needed.

Resolving vectors $\mathbf{v}_s^s$ and $\mathbf{v}_r^s$ into their $dq$ components,

## DYNAMIC MODEL OF THE INDUCTION MOTOR

$$\begin{bmatrix} v_{ds}^s \\ v_{qs}^s \\ v_{dr}^s \\ v_{qr}^s \end{bmatrix} = \begin{bmatrix} R_s & 0 & 0 & 0 \\ 0 & R_s & 0 & 0 \\ 0 & \omega_o L_m & R_r & \omega_o L_r \\ -\omega_o L_m & 0 & -\omega_o L_r & R_r \end{bmatrix} \begin{bmatrix} i_{ds}^s \\ i_{qs}^s \\ i_{dr}^s \\ i_{qr}^s \end{bmatrix}$$

$$+ \begin{bmatrix} L_s & 0 & L_m & 0 \\ 0 & L_s & 0 & L_m \\ L_m & 0 & L_r & 0 \\ 0 & L_m & 0 & L_r \end{bmatrix} \cdot \frac{d}{dt} \begin{bmatrix} i_{ds}^a \\ i_{qs}^s \\ i_{dr}^s \\ i_{qr}^s \end{bmatrix} . \quad (1.34)$$

Using matrix inversion, the voltage equation (1.34) can be rearranged into a current equation

$$\frac{d}{dt} \begin{bmatrix} i_{ds}^s \\ i_{qs}^s \\ i_{dr}^s \\ i_{qr}^s \end{bmatrix} = \frac{1}{L_\sigma^2} \left( \begin{bmatrix} L_r & 0 & -L_m & 0 \\ 0 & L_r & 0 & -L_m \\ -L_m & 0 & L_s & 0 \\ 0 & -L_m & 0 & L_s \end{bmatrix} \begin{bmatrix} v_{ds}^s \\ v_{qs}^s \\ v_{dr}^s \\ v_{qr}^s \end{bmatrix} \right.$$

$$+ \begin{bmatrix} -R_s L_r & \omega_o L_m^2 & R_r L_m & \omega_o L_r L_m \\ -\omega_o L_m^2 & -R_s L_r & -\omega_o L_r L_m & R_r L_m \\ R_s L_m & -\omega_o L_s L_m & -R_r L_s & -\omega_o L_s L_m \\ \omega_o L_s L_m & R_s L_m & \omega_o L_s L_r & -R_r L_s \end{bmatrix} \begin{bmatrix} i_{ds}^s \\ i_{qs}^s \\ i_{dr}^r \\ i_{qr}^s \end{bmatrix} \right)$$

$$(1.35)$$

where

$$L_\sigma = \sqrt{L_s L_r - L_m^2} . \quad (1.36)$$

Eq. (1.35) represents a set of four ordinary differential equations which, along with the $abc \rightarrow dq$ and $dq \rightarrow abc$ transformation equations, allow the current response of the motor to a given supply (stator) voltage to be determined, if the rotor speed, $\omega_o$, is known. This speed, however, results from the interaction between the motor and its mechanical load. Therefore, to model the whole drive system additional equations are needed, namely the load equation, control system equation, and torque equation of the motor. The latter will be derived in the next section.

From here on, $v_{dr}^s$ and $v_{qr}^s$ will be assumed zero because of the shorted rotor winding normally used in induction motors.

## 1.4. Torque Equation

In the primitive two-phase motor, three-phase stator windings of a realistic three-phase machine have been replaced by two perpendicular stationary coils. For analytical purposes, one further step can be made by imagining that the rotating mmf vector $\mathscr{F}_s$ is produced by a fictitious single coil revolving with angular speed $\omega$. Similarly, the rotor vectors can be attributed to a fictitious single coil, revolving with the same speed because of the combined rotation of the rotor vectors round and with the rotor (Eq. (1.6)).

The fictitious revolving stator and rotor coils carrying currents $i_s$ and $i_r$, respectively, are shown in Figure 1.8 in the $d$-$q$ coordinate system (stator reference frame) at a certain instant of time. Both coils consist of a single turn, hence their mmf's equal the conducted currents, i.e., $\mathscr{F}_s = i_s^a$ and $\mathscr{F}_r = i_r^a$. Current vectors $i_s^a$ and $i_r^a$ are displaced in space by angle $\delta$. Dotted lines represent lines of magnetic flux $\Phi_{rs}$ produced by the stator coil and linking the rotor coil. The radius of the rotor coil is denoted by $r$.

The electrodynamic force $F_1$ generated in each of the rotor coil sides is perpendicular to the lines of flux. The torque-producing component $F$ of this force, perpendicular to the coil radius, is

$$F = F_1 \sin(\delta) . \tag{1.37}$$

Since

$$F_1 = B_{rs} l i_r \tag{1.38}$$

where $B_{rs}$ is the flux density of the magnetic field in question and $l$ is the length of a coil side subjected to this field, then the torque $T'$ produced by current $i_r$ in both the coil sides is

$$T' = 2rB_{rs}li_r\sin(\delta). \qquad (1.39)$$

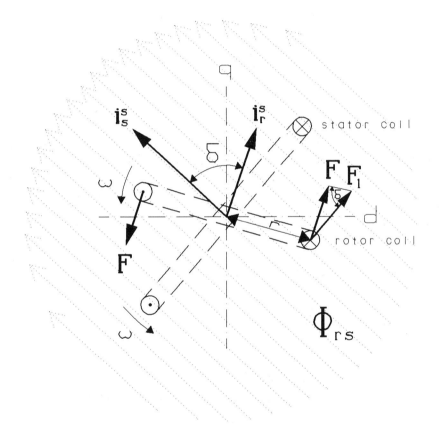

**Figure 1.8.** Fictitious revolving stator and rotor coils.

The flux density, $B_{rs}$, can be expressed in terms of the flux $\Phi_{rs}$ which, again, because of the single-turn stator coil, equals the corresponding flux linkage $\lambda_{rs}$, as

$$B_{rs} = \frac{\Phi_{rs}}{A_r} = \frac{\lambda_{rs}}{A_r} = \frac{\lambda_{rs}}{2lr} \qquad (1.40)$$

where $A_r$ is the cross-sectional area of the rotor coil. After substitution of Eq. (1.40) in Eq. (1.39),

$$T' = \lambda_{rs} i_r \sin(\delta). \qquad (1.41)$$

For clarity, it should be noted that flux linkage $\lambda_{rs}$, given by

$$\lambda_{rs} = L_m i_s \qquad (1.42)$$

represents only a portion of the total rotor flux linkage $\lambda_r$. The other component of the rotor flux linkage, i.e., the one produced by the rotor current itself, does not contribute to the torque production. The electrodynamic forces resulting from this flux linkage are only pulling the coil sides apart from each other, due to the known phenomenon of repulsion of two parallel conductors carrying currents of opposite polarity, and no force component perpendicular to the coil radius appears. If this were not the case, the stator simply would not be needed since an energized rotor coil could propel itself, or, in general, no magnetic field source external to the armature circuit would be necessary in electric motors.

Current vectors $\vec{i}_s^s$ and $\vec{i}_r^s$ from Figure 1.8 are shown again in Figure 1.9 where it can be seen that

$$\begin{aligned}
\sin(\delta) &= \sin(\delta_s - \delta_r) \\
&= \sin(\delta_s)\cos(\delta_r) - \cos(\delta_s)\sin(\delta_r) \\
&= \frac{i_{qs}^s}{i_s} \cdot \frac{i_{dr}^s}{i_r} - \frac{i_{ds}^s}{i_s} \cdot \frac{i_{qr}^s}{i_r} \\
&= \frac{1}{i_s i_r}(i_{qs}^s i_{dr}^s - i_{ds}^s i_{qr}^s).
\end{aligned} \qquad (1.43)$$

Substitution of Eqs. (1.42) and (1.43) in Eq. (1.41) yields

$$T' = L_m(i_{qs}^s i_{dr}^s - i_{ds}^s i_{dr}^s). \qquad (1.44)$$

# DYNAMIC MODEL OF THE INDUCTION MOTOR

At this point, the question arises of how the torque $T$ of a real, $P$-pole, three-phase motor differs from that of the analysed abstract, 2-pole, two-phase primitive model. Firstly, there is the difference in speed of the motors. The $P$-pole motor rotates $P/2$ times slower than the 2-pole one. Consequently, at the same output power, the $P$-pole motor develops a torque which is $P/2$ times greater than that of the primitive machine. Secondly, there is the difference in input power. It has been shown in Section 1.1 that when the primitive motor is supplied with a balanced, sinusoidal voltage, then

$$i_s = \mathcal{I}_s = 1.5 I_{s,\max} = 1.5\sqrt{2} I_s \qquad (1.45)$$

where $I_{s,\max}$ and $I_s$ are the peak and r.m.s. value of the stator current, respectively. Analogously,

$$v_s = 1.5\sqrt{2} V_s \qquad (1.46)$$

where $V_s$ denotes the r.m.s. value of the stator voltage. In effect, the apparent power supplied to the fictitious rotating stator coil in Figure 1.8 is

$$S_s' = v_s i_s = 4.5 V_s I_s. \qquad (1.47)$$

As the apparent power supplied to the real, three-phase motor is

$$S_s = 3 V_s I_s \qquad (1.48)$$

then, assuming that the both machines have equal efficiencies and power factors, the ratio of the output real power of the real motor to that of the primitive one equals 2/3. As a result, the torque developed by the real motor is

$$T = \frac{P}{2} \cdot \frac{2}{3} T' = \frac{P}{3} L_m (i_{qs}^s i_{dr}^s - i_{ds}^s i_{qr}^r). \qquad (1.49)$$

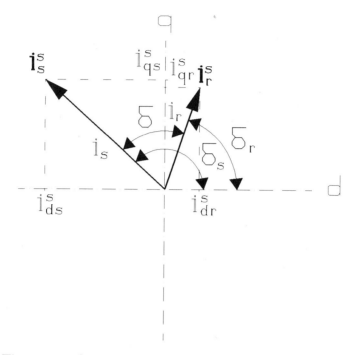

**Figure 1.9.** Current vectors in the stator reference frame.

It can easily be shown that the term in parentheses equals the imaginary part of the product of vector $\mathbf{i}_s^s$ and conjugate vector $\mathbf{i}_r^{s*}$. In effect, another expression for the torque, often encountered in the literature, can be written as

$$T = \frac{P}{3} L_m Im(\mathbf{i}_s^s \mathbf{i}_r^{s*}). \qquad (1.50)$$

## 1.5 Dynamic Equivalent Circuit

For an induction motor with shorted rotor winding, Eqs. (1.26) and (1.27), can be written as

$$\boldsymbol{\lambda}_s^s = L_s \mathbf{i}_s^s + L_m \mathbf{i}_r^s \qquad (1.51)$$

$$\pmb{\lambda}_r^s = L_m \pmb{i}_s^s + L_r \pmb{i}_r^s \qquad (1.52)$$

$$\pmb{v}_s^s = R_s \pmb{i}_s^s + p\pmb{\lambda}_s^s \qquad (1.53)$$

$$0 = R_r \pmb{i}_r^s + p\pmb{\lambda}_r^s - j\omega_o \pmb{\lambda}_r^s. \qquad (1.54)$$

Since

$$L_s = L_{ls} + L_m \qquad (1.55)$$

and

$$L_r = L_{lr} + L_m \qquad (1.56)$$

then

$$\pmb{\lambda}_s^s = L_{ls} \pmb{i}_s^s + L_m (\pmb{i}_s^s + \pmb{i}_r^s) \qquad (1.57)$$

$$\pmb{\lambda}_r^s = L_{lr} \pmb{i}_r^s + L_m (\pmb{i}_s^s + \pmb{i}_r^s). \qquad (1.58)$$

Eqs. (1.53), (1.54), (1.57), and (1.58) describe the circuit shown in Figure 1.10. Its structure resembles that of the classic, per-phase equivalent circuit of an induction motor in the steady state, shown in Figure 1.11, were motor quantities are represented as phasors, $\hat{I}_m$ denoting the magnetizing current and $s$ the slip of the motor, equal

$$s = \frac{\omega_r}{\omega}. \qquad (1.59)$$

The similarity between the vector-based dynamic equivalent circuit and the phasor-based steady-state equivalent circuit is, of course, not a coincidence. To show that the latter circuit represents a special case of the former, the relation between the vector and phasor quantities of an induction motor will first be explained.

For analytical purposes, it can be assumed that the coordinates of the abstract, complex phasor plane are aligned with the *dq* coordinates of the

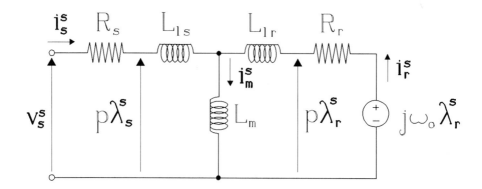

**Figure 1.10.** Dynamic equivalent circuit of induction motor.

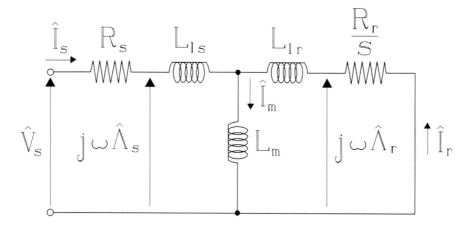

**Figure 1.11.** Steady-state equivalent circuit of induction motor.

physical, vector plane. Then, considering, for instance, the stator-voltage vector, $v_s^s$, resulting from a balanced, sinusoidal three-phase supply, and the corresponding phasor, $\hat{V}_s$, the following equation applies:

$$v_s^s = 1.5\sqrt{2}\hat{V}_s e^{j\omega t}. \tag{1.60}$$

The individual terms on the right side of Eq. (1.60) can be explained as follows:

(1) The coefficient 1.5 results from the conversion of the three-phase real motor into the two-phase equivalent model (phasor $\hat{V}_s$ represents the stator voltage per single phase of the three-phase motor).

(2) The coefficient √2 results from the convention that since vectors are calculated from time functions of physical quantities of the motor, then their peak values are used. In contrast, in the steady-state analysis of power apparatus, phasor magnitudes are taken as r.m.s. values of those quantities.

(3) The $e^{j\omega t}$ term indicates rotation of the vector in the physical space of the motor.

It must be stressed again that Eq. (1.60) and its equivalents for other quantities of the motor are valid only when these quantities are balanced and sinusoidal, so that they can be expressed by phasors. Such a situation occurs when the motor operates in the steady state while supplied from a balanced three-phase source of a.c. voltage.

Eqs. (1.26) and (1.27), can be re-written as

$$1.5\sqrt{2}\hat{\Lambda}_s e^{j\omega t} = 1.5\sqrt{2}L_{ls}\hat{I}_s e^{j\omega t} \\ + L_m(1.5\sqrt{2}\hat{I}_s e^{j\omega t} \\ + 1.5\sqrt{2}\hat{I}_r e^{j\omega t}) \quad (1.61)$$

$$1.5\sqrt{2}\hat{\Lambda}_r e^{j\omega t} = 1.5\sqrt{2}L_{lr}\hat{I}_r e^{j\omega t} \\ + L_m(1.5\sqrt{2}\hat{I}_s e^{j\omega t} \\ + 1.5\sqrt{2}\hat{I}_r e^{j\omega t}) \quad (1.62)$$

$$1.5\sqrt{2}\hat{V}_s = 1.5\sqrt{2}R_s\hat{I}_s e^{j\omega t} \\ + \frac{d}{dt}(1.5\sqrt{2}\hat{\Lambda}_s e^{j\omega t}) \quad (1.63)$$

$$0 = 1.5\sqrt{2}R_r\hat{I}_r e^{j\omega t}$$
$$+ \frac{d}{dt}(1.5\sqrt{2}\hat{\Lambda}_r e^{j\omega t}) \qquad (1.64)$$
$$- j1.5\sqrt{2}\omega_o \hat{\Lambda}_r e^{j\omega t}.$$

Taking into account that

$$\frac{d}{dt}(1.5\sqrt{2}\hat{\Lambda}_s e^{j\omega t}) = j1.5\sqrt{2}\omega\hat{\Lambda}_s e^{j\omega t} \qquad (1.65)$$

$$\frac{d}{dt}(1.5\sqrt{2}\hat{\Lambda}_r e^{j\omega t}) = j1.5\sqrt{2}\omega\hat{\Lambda}_r e^{j\omega t} \qquad (1.66)$$

and reducing these by the $1.5\sqrt{2}e^{j\omega t}$ term, Eqs. (1.61) to (1.64) can be rewritten as

$$\hat{\Lambda}_s = L_{ls}\hat{I}_s + L_m(\hat{I}_s + \hat{I}_r) \qquad (1.67)$$

$$\hat{\Lambda}_r = L_{lr}\hat{I}_r + L_m(\hat{I}_s + \hat{I}_r) \qquad (1.68)$$

$$\hat{V}_s = R_r\hat{I}_s + j\omega\hat{\Lambda}_s \qquad (1.69)$$

$$0 = R_r\hat{I}_r + j(\omega - \omega_o)\hat{\Lambda}_r \qquad (1.70)$$

The term in parentheses in the last equation represents the rotor radian frequency $\omega_r$, i.e., the frequency of rotor currents in a two-pole motor. Multiplying both sides of Eq. (1.70) by $\omega/\omega_r$, gives

$$\frac{\omega}{\omega_r}R_r\hat{I}_r + j\omega\hat{\Lambda}_r = \frac{R_r}{S}\hat{I}_r + j\omega\hat{\Lambda}_r = 0. \qquad (1.71)$$

# DYNAMIC MODEL OF THE INDUCTION MOTOR

It can be seen that Eqs. (1.67) to (1.69), and (1.71) describe the steady-state, per-phase equivalent circuit of the induction motor, shown in Figure 1.11. The torque developed by the motor can be calculated from the steady-state, phasor version of Eq. (1.50) as

$$T = 1.5 P L_m \text{Im}(\hat{I}_s \hat{I}_r^*) . \quad (1.72)$$

where $\hat{I}_r^*$ is the conjugate phasor of rotor current. Finally, Eqs. (1.69) and (1.71) can be expressed in a matrix form, analogous to that of Eq. (1.27), as

$$\begin{bmatrix} \hat{V}_s \\ 0 \end{bmatrix} = \begin{bmatrix} R_s + j\omega L_s & j\omega L_m \\ j\omega L_m & \dfrac{R_r}{s} + j\omega L_r \end{bmatrix} \begin{bmatrix} \hat{I}_s \\ \hat{I}_r \end{bmatrix} \quad (1.73)$$

which, along with Eq. (1.72), allows analysis of the steady-state operation of an induction motor to be carried out.

## 1.6 Direct and Inverse Stator to Excitation Reference Frame Transformations

Eq. (1.60) can be rewritten in the following form:

$$\begin{aligned} v_{ds}^s + j v_{qs}^s &= 1.5\sqrt{2} V_s e^{j\varphi_s} e^{j\omega t} \\ &= 1.5\sqrt{2} V_s \cos(\omega t + \varphi_s) \\ &\quad + j 1.5\sqrt{2} V_s \sin(\omega t + \varphi_s) \end{aligned} \quad (1.74)$$

where $V_s$ and $\varphi_s$ are the magnitude and phase of phasor $\hat{V}_s$, respectively. Consequently,

$$v_{ds}^s = 1.5\sqrt{2} V_s \cos(\omega t + \varphi_s) \quad (1.75)$$

$$v_{qs}^s = 1.5\sqrt{2} V_s \sin(\omega t + \varphi_s) \quad (1.76)$$

which means that under the sinusoidal steady state conditions the $dq$

components of the vector quantities have sinusoidal, a.c. waveforms. This is so because of the mentioned physical interpretation of the $dq$ components as phase quantities of an equivalent two-phase machine.

A.c. quantities are somewhat inconvenient for control purposes. For instance, control systems are usually represented by block diagrams in which the variables are time-varying d.c. signals. Therefore, another transformation will be introduced which allows conversion of the a.c. $dq$ components of the motor vectors into d.c. variables.

The transformation in question involves the so-called excitation reference frame $(D,Q)$ which, in contrast to the stationary, stator reference frame $(d,q)$, rotates with the angular velocity $\omega$ in the same direction as does the stator mmf vector $\mathcal{F}_s$. As a result, in the steady state, coordinates of the motor vectors in the new reference frame do not vary in time. This is illustrated in Figures 1.12 and 1.13 which show the stator mmf vector in both reference frames at two different instants of time differing by one sixth of the cycle of $\omega$. Superscript $e$ and upper-case subscripts are used for the vectors and their components in the excitation frame. The same convention will be employed with respect to the other vector quantities of the motor.

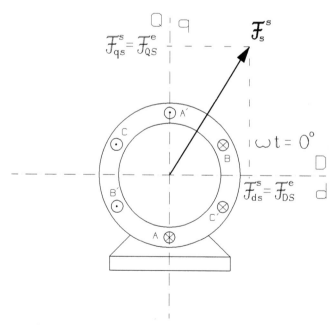

**Figure 1.12.** Stator mmf vector in the stator and excitation reference frames: initial instant of time.

# DYNAMIC MODEL OF THE INDUCTION MOTOR 27

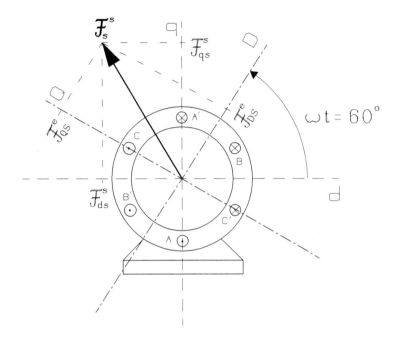

**Figure 1.13.** Stator mmf vector in the stator and excitation reference frames: one sixth of the cycle of $\omega$ later.

Considering, for instance, the stator voltage, the relation between vector $v_s^s$ in the stator frame and vector $v_s^e$ in the excitation frame is

$$v_s^s = v_s^e e^{j\omega t} \qquad (1.77)$$

or

$$v_s^e = v_s^s e^{-j\omega t}. \qquad (1.78)$$

Resolution of both vectors in Eq. (1.78) yields

$$\begin{aligned}
v_{DS}^e + jv_{QS}^e &= (v_{ds}^s + jv_{qs}^s)[\cos(-\omega t) + j\sin(-\omega t)] \\
&= v_{ds}^s \cos(\omega t) + v_{qs}^s \sin(\omega t) \\
&\quad + j[-v_{ds}^s \sin(\omega t) + v_{qs}^s \cos(\omega t)].
\end{aligned} \qquad (1.79)$$

Hence, the transformation $dq \rightarrow DQ$ from the stator reference frame to the excitation frame can be expressed as

$$\begin{bmatrix} v_{DS}^e \\ v_{QS}^e \end{bmatrix} = \begin{bmatrix} \cos(\omega t) & \sin(\omega t) \\ -\sin(\omega t) & \cos(\omega t) \end{bmatrix} \begin{bmatrix} v_{ds}^s \\ v_{qs}^s \end{bmatrix} \qquad (1.80)$$

while the inverse, $DQ \rightarrow dq$ transformation is

$$\begin{bmatrix} v_{ds}^s \\ v_{qs}^s \end{bmatrix} = \begin{bmatrix} \cos(\omega t) & -\sin(\omega t) \\ \sin(\omega t) & \cos(\omega t) \end{bmatrix} \begin{bmatrix} v_{DS}^e \\ v_{QS}^e \end{bmatrix}. \qquad (1.81)$$

## 1.7 Motor Equations in Excitation Reference Frame

The voltage and current equations of the induction motor derived in Section 1.3 in the stator reference frame can easily be converted into those in the excitation frame. Starting with Eq. (1.16), this can be rewritten as

$$\boldsymbol{v}_s^e e^{j\omega t} = R_s \boldsymbol{1}_s^e e^{j\omega t} + p\boldsymbol{\lambda}_s^e e^{j\omega t}. \qquad (1.82)$$

Since $p$ denotes the differentiation operator, then

$$\begin{aligned} p\boldsymbol{\lambda}_s^e e^{j\omega t} &= \frac{d\boldsymbol{\lambda}_s^e}{dt} e^{j\omega t} + j\omega e^{j\omega t} \boldsymbol{\lambda}_s^e \\ &= (p+j\omega) \boldsymbol{\lambda}_s^e e^{j\omega t}. \end{aligned} \qquad (1.83)$$

Substituting Eq. (1.83) in Eq. (1.82) and cancelling the $e^{j\omega t}$ term gives

$$\boldsymbol{v}_s^e = R_s \boldsymbol{1}_s^e + (p+j\omega) \boldsymbol{\lambda}_s^e. \qquad (1.84)$$

# DYNAMIC MODEL OF THE INDUCTION MOTOR

It can be seen that Eq. (1.84) could have been obtained by substituting $(p + j\omega)$ for $p$ in Eq. (1.24). Similarly, Eq. (1.25) yields

$$\begin{aligned} \mathbf{v}_R^e &= R_r \mathbf{1}_R^e + (p + j\omega - j\omega_o) \boldsymbol{\lambda}_R^e \\ &= R_r \mathbf{1}_R^e + (p + j\omega_r) \boldsymbol{\lambda}_R^e. \end{aligned} \quad (1.85)$$

As

$$\boldsymbol{\lambda}_S^e = L_s \mathbf{1}_S^e + L_m \mathbf{1}_R^e \quad (1.86)$$

$$\boldsymbol{\lambda}_R^e = L_m \mathbf{1}_S^e + L_r \mathbf{1}_R^e \quad (1.87)$$

then

$$\begin{bmatrix} \mathbf{v}_S^e \\ \mathbf{v}_R^e \end{bmatrix} = \begin{bmatrix} R_s + (p+j\omega)L_s & (p+j\omega)L_m \\ (p+j\omega_r)L_m & R_r + (p+j\omega_r)L_r \end{bmatrix} \begin{bmatrix} \mathbf{1}_S^e \\ \mathbf{1}_R^e \end{bmatrix} \quad (1.88)$$

or

$$\begin{bmatrix} v_{DS}^e \\ v_{QS}^e \\ v_{DR}^e \\ v_{QR}^e \end{bmatrix} = \begin{bmatrix} R_s + pL_s & -\omega L_s & pL_m & -\omega_r L_m \\ \omega L_s & R_s + pL_s & \omega_r L_m & pL_m \\ pL_m & -\omega_r L_m & R_r + pL_r & -\omega_r L_r \\ \omega_r L_m & pL_m & \omega_r L_r & pL_r \end{bmatrix} \begin{bmatrix} i_{DS}^e \\ i_{QS}^e \\ i_{DR}^e \\ i_{QR}^e \end{bmatrix} \cdot (1.89)$$

Again, the rotor voltage vector is normally assumed zero because of the shorted rotor winding, i.e., $v_{DS}^e = v_{QS}^e = 0$.

The torque equation in the excitation reference frame is similar to that in the stator frame as the terms $e^{j\omega t}$ and $e^{-j\omega t}$ cancel, i.e.,

$$T = \frac{P}{3} L_m (i^e_{QS} i^e_{DR} - i^e_{DS} i^e_{QR})$$
$$= \frac{P}{3} L_m Im(\mathbf{1}^e_S \mathbf{1}^{e*}_R) \,. \tag{1.90}$$

## 1.8 Examples and Simulations

Listed below are parameters of the motor used in all the examples and simulations in this book. The MATLAB program from The MathWorks Inc. was used in numerical examples, but any other software handling complex numbers and matrices can be employed for similar computations. The simulations have been performed using the ACSL dynamic simulation software package from Mitchell & Gauthier Associates.

### Parameters of the Example Motor

- Type: three-phase, wye-connected, squirrel-cage induction motor
- Rated power: $HP_{rat}$ = 10 hp (7.46 kW)
- Rated stator voltage: $V_{s,rat}$ = 220 V
- Rated frequency: $f_{rat}$ = 60 Hz
- Rated speed: $n_{M,rat}$ = 1164 rpm
- Number of poles: $P$ = 6
- Stator resistance: $R_s$ = 0.294 Ω/ph
- Stator leakage reactance: $X_{ls}$ = 0.524 Ω/ph
- Rotor resistance referred to stator: $R_r$ = 0.156 Ω/ph
- Rotor leakage reactance referred to stator: $X_{lr}$ = 0.279 Ω/ph
- Magnetizing reactance: $X_m$ = 15.457 Ω/ph
- Mass moment of inertia of the rotor: $J_M$ = 0.4 kg m².

### Example 1.1.

The motor operates in the steady state under the rated operating conditions. Calculate phasors of the stator and rotor currents and the developed torque from:

(a) solution of the steady-state, per-phase equivalent circuit,

(b) equations (1.72) and (1.73).

# DYNAMIC MODEL OF THE INDUCTION MOTOR

## Solution

(a) The steady-state, per-phase equivalent circuit of the motor, corresponding to that in Figure 1.11, is shown in Figure 1.14. Phasor $\hat{V}_s$ is taken as a reference phasor and, due to the wye connection of stator windings, is given by

$$\hat{V}_s = \frac{220}{\sqrt{3}} = 127 \ V/ph.$$

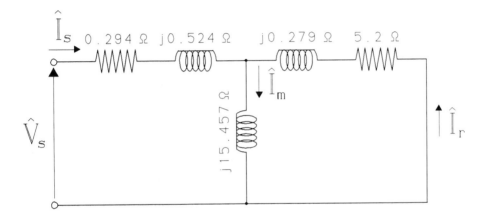

**Figure 1.14.** Steady-state, per-phase equivalent circuit of the example motor.

The driving-point impedance, $\hat{Z}$, of the circuit is

$$\hat{Z} = 0.294 + j0.524 + \frac{j15.427(5.2 + j0.279)}{j15.427 + 5.2 + j0.279}$$
$$= 5.335 \angle 25.5° \ \Omega/ph.$$

Hence, phasors $\hat{I}_s$ and $\hat{I}_r$ can be calculated as

$$\hat{I}_s = \frac{\hat{V}_s}{\hat{Z}} = \frac{127}{5.355 \angle 25.5°} = \mathbf{23.8 \angle -25.5° \ A/ph}$$

and

$$\hat{I}_r = -\hat{I}_s \frac{j15.427}{j15.427+5.2+j0.279}$$
$$= -23.8\angle-25.5° \times 0.933\angle 18.3°$$
$$= \mathbf{22.2\angle 172.8° \ A/ph}.$$

The total (three-phase) electrical power supplied to the rotor is

$$P_{elec} = 3I_r^2 \frac{R_r}{s} = 3\times 22.2^2 \times 5.2 = 7,688 \ W.$$

Subtracting the total power loss in the rotor winding, the mechanical power is obtained as

$$P_{mech} = P_{elec} - 3I_r^2 R_r$$
$$= 7,688 - 3\times 22.2^2 \times 0.156 = 7,457 \ W.$$

The angular velocity, $\omega_M$, of the rotor is

$$\omega_M = \frac{2\pi}{60} n_{M,rat} = \frac{\pi}{30}\times 1200 = 121.9 \ rad/sec.$$

Hence, the torque developed by the motor can be determined by dividing the mechanical power by the speed of the rotor as

$$T = \frac{P_{mech}}{\omega_M} = \frac{7,457}{121.9} = \mathbf{61.2 \ N \ m}.$$

(b) The stator and rotor reactances are

$$X_s = X_{ls} + X_m = 0.524 + 15.457 = 15.981 \ \Omega/ph$$

and

$$X_r = X_{lr} + X_m = 0.279 + 15.457 = 15.736 \ \Omega/ph$$

Hence, Eq. (1.73) is

$$\begin{bmatrix} 127 \\ 0 \end{bmatrix} = \begin{bmatrix} 0.294+j15.981 & j15.457 \\ j15.457 & 5.2+j15.736 \end{bmatrix} \begin{bmatrix} \hat{I}_s \\ \hat{I}_r \end{bmatrix}.$$

Consequently,

$$\begin{bmatrix} \hat{I}_s \\ \hat{I}_r \end{bmatrix} = \begin{bmatrix} 0.294+j15.981 & j15.457 \\ j15.457 & 5.2+j15.736 \end{bmatrix}^{-1} \begin{bmatrix} 127 \\ 0 \end{bmatrix}$$

$$= \begin{bmatrix} 21.49-j10.23 \\ -22.03+j2.77 \end{bmatrix} = \begin{bmatrix} \mathbf{23.8\angle-25.5°} \\ \mathbf{22.2\angle172.8°} \end{bmatrix} \ \pmb{A/ph}.$$

From Eq. (1.72),

$$T = 1.5 \times 6 \times 0.041 \times Im(23.8\angle-25.5° \times 22.2\angle-172.8°)$$
$$= 195\sin(-198.3°) = \mathbf{61.2 \ N \ m}.$$

As expected, both the approaches used have yielded identical results.

### Example 1.2.

Use the matrix equation (1.27) to determine the stator and rotor current vectors in the stator reference frame, and calculate the torque of the motor under the same operating conditions as those in Example 1.1.

### *Solution*

The synchronous speed, $\omega$, equals $2\pi f = 2\pi \times 60 = 377$ rad/sec, and the motor inductances are:

$$L_s = \frac{X_s}{\omega} = \frac{15.981}{377} = 0.0424 \; H/ph$$

$$L_r = \frac{X_r}{\omega} = \frac{15.736}{377} = 0.0417 \; H/ph$$

$$L_m = \frac{X_m}{\omega} = \frac{15.457}{377} = 0.041 \; H/ph$$

$$\omega_o = (1-s)\omega = (1-0.03) \times 377 = 365.7 \; rad/sec.$$

The vector of stator voltage, calculated from Eq. (1.60), is

$$\mathbf{v}_s^s = 1.5\sqrt{2} \times 127 e^{j377t} = 269.4 e^{j377t} \; V$$

and Eq. (1.27) can be written as

$$\begin{bmatrix} 269.4 e^{j377t} \\ 0 \end{bmatrix} =$$
$$= \begin{bmatrix} 0.294 + 0.0424p & 0.041p \\ (p-j365.7)0.041 & 0.156 + (p-j365.7)0.0417 \end{bmatrix} \begin{bmatrix} \mathbf{1}_s^s \\ \mathbf{1}_r^s \end{bmatrix}.$$

Since the motor operates in the steady state, the current vectors do not change their magnitudes, while rotating with the constant speed $\omega$. Hence,

$$\mathbf{1}_s^s = I_s e^{j(377t+\delta_s)}$$

and

$$\mathbf{1}_r^s = I_r e^{j(377t+\delta_r)}.$$

# DYNAMIC MODEL OF THE INDUCTION MOTOR

The differentiation operator, $p$, in the matrix equation can be replaced by $j\omega = j377$ because

$$p\mathbf{1}_s^s = \frac{d}{dt}I_s e^{j(377t+\delta_s)}$$
$$= j377 I_s e^{j(377t+\delta_s)} = j377 \mathbf{1}_s^s$$

and

$$p\mathbf{1}_r^s = j377 \mathbf{1}_r^s.$$

After this substitution, the matrix equation becomes

$$\begin{bmatrix} 269.4 e^{j377t} \\ 0 \end{bmatrix} = \begin{bmatrix} 0.294+j15.981 & j15.457 \\ j0.463 & 0.156+j0.472 \end{bmatrix} \begin{bmatrix} \mathbf{1}_s^s \\ \mathbf{1}_r^s \end{bmatrix}$$

or

$$\begin{bmatrix} \mathbf{1}_s^s \\ \mathbf{1}_r^s \end{bmatrix} = \begin{bmatrix} 0.294+j15.981 & j15.457 \\ j0.463 & 0.156+j0.472 \end{bmatrix}^{-1} \begin{bmatrix} 269.4 e^{j377t} \\ 0 \end{bmatrix}$$
$$= \begin{bmatrix} 50.5 e^{j(377t-25.5°)} \\ 47.1 e^{j(377t+172.8°)} \end{bmatrix} \text{A.}$$

The solution obtained can be confirmed using the vector/phasor relation (1.60) and phasors $\hat{I}_s$ and $\hat{I}_r$ found in Example 1.1. Indeed,

$$\begin{bmatrix} \mathbf{1}_s^s \\ \mathbf{1}_r^s \end{bmatrix} = 1.5\sqrt{2} e^{j377t} \begin{bmatrix} \hat{I}_s \\ \hat{I}_r \end{bmatrix} = 1.5\sqrt{2} e^{j377t} \begin{bmatrix} 23.8 e^{-j25.5°} \\ 22.2 e^{j172.8°} \end{bmatrix}$$
$$= \begin{bmatrix} 50.5 e^{j(377t-25.5°)} \\ 47.1 e^{j(377t+172.8°)} \end{bmatrix} \text{A.}$$

From Eq. (1.50), the torque developed in the motor is

$$T = \frac{6}{3} \times 0.041 \, Im[50.5 e^{j(377t-25.5°)} \times 47.1 e^{-j(377t+172.8°)}]$$
$$= \frac{6}{3} \times 0.041 \times 50.5 \times 47.1 \sin(-25.5° - 172.8°)$$
$$= \mathbf{61.2 \ N \ m}.$$

The same result has been obtained in Example 1.1.

### Simulation 1.1. Direct-On-Line Starting

To illustrate the application of the dynamic model of the induction motor to the investigation of transient operation of the motor, simulation of direct-on-line starting is demonstrated. At the initial instant of time, $t = 0$, the motor, previously de-energized and stalled, is directly connected to the power line at the full rated voltage of 220 V and frequency of 60 Hz. The load torque, $T_L$, is assumed to be 50% of the rated torque of the motor, and independent of speed. The mass moment of inertia, $J_L$, of the load equals that of the motor.

The simulation consists of numerical solution of the following set of simultaneous equations:

(1) current equation (1.35),
(2) torque equation (1.49),
(3) load equation

$$\frac{d\omega_M}{dt} = \frac{T - T_L}{J_M + J_L} \qquad (1.91)$$

where $\omega_M$ is the angular velocity of the rotor. The relation between $\omega_o$, appearing in the current equation, and $\omega_M$ is

$$\omega_o = \frac{P}{2} \omega_M. \qquad (1.92)$$

The $abc \to dq$ and $dq \to abc$ transformations must be performed at each step of the numerical procedure to convert the three-phase voltages and currents into these in the stator reference frame, used in the current and torque equations.

The developed torque of the motor and speed of the system are shown in Figure 1.15. Strong oscillations of the torque occur in the initial stage of the starting process. Later, the oscillations disappear, the torque passes through its peak value, and stabilizes at the steady-state level of the load torque. Within 1.7 sec, the motor accelerates to the steady-state speed of 1183 rpm.

Waveforms of rotor and stator currents, $i_{as}$ and $i_{ar}$, are shown in Figure 1.16. They are very similar, due to the fact that the rotor current displayed is not the actual current in the squirrel cage of the rotor, but the current referred to the stator. Within the major part of the acceleration period, in spite of the relatively light load of the motor, the currents are up to six times higher than their steady-state values.

These results confirm the well known disadvantages of direct-on-line starting of induction motors. High torque oscillations, particularly in frequently started drive systems, reduce the life span of mechanical couplings and gears, while the high starting current, especially with heavy loads, may cause overheating and failure of the motor. Therefore, in many instances, assisted starting is performed by applying reduced voltage to the stator (using an autotransformer or a wye/delta switch) or employing an adjustable frequency and voltage supply source (inverter).

### Simulation 1.2. Uncontrolled Motor - Load Changes

The purpose of this simulation is to illustrate the difference between motor quantities in the stator and excitation reference frames. The same drive system as that in Simulation 1.1 undergoes sudden changes in the load torque: at $t = 1$ sec, the load increases from 50% to 150% of the rated torque, $T_{rat}$, then at $t = 3$ sec, decreases to 100% of $T_{rat}$.

Variations of motor torque and speed are shown in Figure 1.17. Following the load changes, the speed, initially at 1183 rpm, drops to 1141 rpm, then increases to 1164 rpm. The stator and rotor currents in the stator reference frame, shown in Figures 1.18 and 1.19, change their amplitudes in proportion to the torque. The currents have a.c. waveforms with a frequency equal to that of the supply voltage, i.e., 60 Hz.

Corresponding waveforms of the motor currents in the excitation frame are shown in Figures 1.20 and 1.21. These are seen to be time-varying d.c. signals, which are easier to analyse and manipulate in control schemes than the a.c. signals in the stator reference frame.

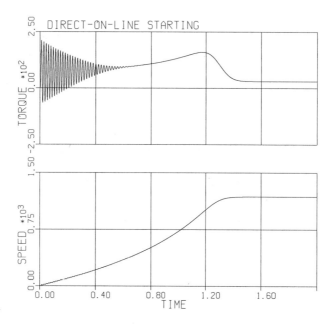

**Figure 1.15.** Torque and speed of the motor with direct-on-line starting.

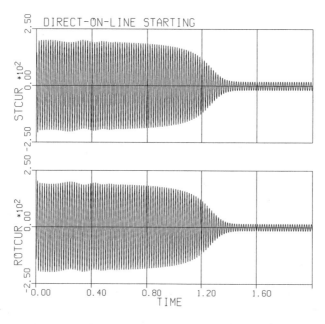

**Figure 1.16.** Stator and rotor current waveforms of the motor with direct-on-line starting.

# DYNAMIC MODEL OF THE INDUCTION MOTOR

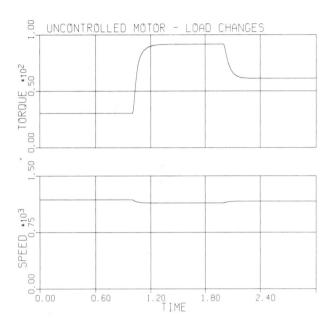

**Figure 1.17.** Torque and speed of the motor with rapid load changes.

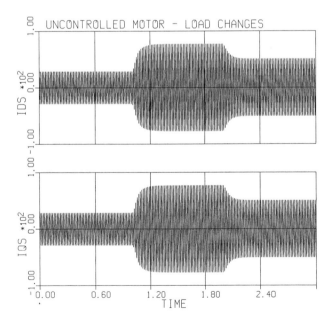

**Figure 1.18.** Stator current waveforms of the motor in the stator reference frame.

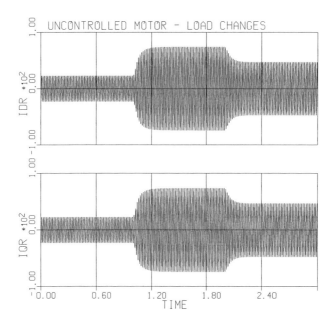

**Figure 1.19.** Rotor current waveforms of the motor in the stator reference frame.

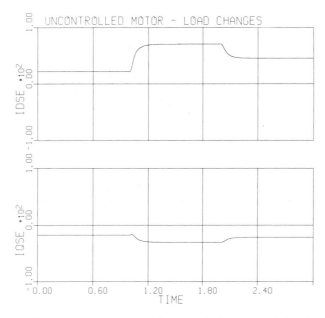

**Figure 1.20.** Stator current waveforms of the motor in the excitation frame.

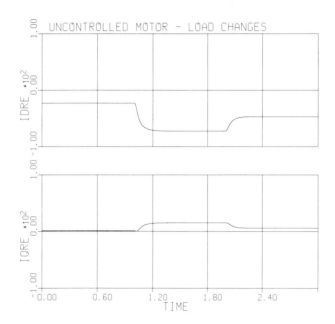

**Figure 1.21.** Rotor current waveforms of the motor in the excitation frame.

# Chapter 2

## SCALAR CONTROL OF INDUCTION MOTORS

A majority of the existing variable speed drive systems with induction motors are low-performance drives in which the adjusted variables are the magnitude and frequency of either the voltage or current supplied to the stator. This allows control of the steady-state speed or torque of the motor, while the magnetic field in the motor is kept at a constant, desired level. This type of control is usually referred to as scalar control, since the controlled stator voltage or current are assumed sinusoidal, with only the magnitude and frequency adjusted, and with no concern for the spatial position (phase) of the corresponding vector quantities. In contrast, vector control schemes, based on the Field Orientation Principle, provide adjustment of the magnitude and phase of the vector quantities of the motor. In vector-controlled, high-performance drive systems, the instantaneous torque and magnetic field of the motor are regulated, both in the steady-state and under transient operating conditions. Scalar control systems are simpler than those employing vector control, but the superiority of the latter from the point of view of dynamic performance of the drive is unqestionable.

Possibly the most common scalar technique used in practice is that of Constant Volts/Hertz (CVH), so named because the magnitude of stator voltage is adjusted in proportion to the frequency, in order to maintain an approximately constant stator flux in the motor. The CHV method consists, in essence, of controlling the speed of the rotating magnetic field of the stator by changing the supply frequency. The torque developed in the motor depends only on the slip speed, i.e., the difference in speeds between the field and the rotor. The control system is simple, since only speed feedback is required. Another scalar control method uses the Torque Control (TC) technique, in which the magnitude and frequency of the stator currents are adjusted so that the steady-state torque is controlled, while the magnetic field is again maintained constant. In this case, the speed feedback has to be supplemented by a current feedback which makes the TC system somewhat more complicated than that of the CVH method.

In this chapter, the principles of scalar control of induction motors are explained in detail, in order to provide a background for the more advanced concept of vector control.

## 2.1 The Γ Equivalent Circuit of an Induction Motor

The per-phase equivalent circuit of an induction motor in the steady state, shown in Figure 1.11, can be called a T circuit, due to the T-shape arrangement of the inductive elements. Parameters of the circuit can be calculated from the no-load and blocked-rotor tests. The no-load test allows determination of the mutual inductance, $L_m$, while the blocked-rotor test yields the combined values of the stator and rotor resistances, $R_s + R_r$, and the stator and rotor leakage inductances, $L_{ls} + L_{lr}$. The resistances can be evaluated separately by an additional, direct measurement of the easily accessible stator resistance. Separation of the leakage inductances is, however, impossible and can only be done using approximate rules of estimation, such as an equal split, proportionality to the corresponding resistances, or empirical relations for a given class of machines.

Another approach consists in transformation of the T equivalent circuit to the so-called Γ circuit which includes only two inductive elements. Vector quantities of the rotor portion of the new circuit will be denoted by upper-case subscripts "$R$", and they are assumed to be linearly related to the corresponding quantities of the T circuit. Specifically,

$$\begin{bmatrix} \mathbf{i}_R^s \\ \boldsymbol{\lambda}_R^s \end{bmatrix} = \begin{bmatrix} \gamma^{-1} & \\ & \gamma \end{bmatrix} \begin{bmatrix} \mathbf{i}_r^s \\ \boldsymbol{\lambda}_r^s \end{bmatrix} \qquad (2.1)$$

or

$$\begin{bmatrix} \mathbf{i}_r^s \\ \boldsymbol{\lambda}_r^s \end{bmatrix} = \begin{bmatrix} \gamma & \\ & \gamma^{-1} \end{bmatrix} \begin{bmatrix} \mathbf{i}_R^s \\ \boldsymbol{\lambda}_R^s \end{bmatrix} \qquad (2.2)$$

where

$$\gamma \equiv \frac{L_s}{L_m}. \qquad (2.3)$$

Substituting Eq. (2.2) in Eqs. (1.25) and (1.26), rearranging, and assuming, as usual, a shorted rotor winding, the following equations are obtained:

$$R_R \mathbf{1}_R^s + (p - j\omega_o)\boldsymbol{\lambda}_R^s = 0 \qquad (2.4)$$

$$\boldsymbol{\lambda}_s^s = L_M(\mathbf{1}_s^s + \mathbf{1}_R^s) \qquad (2.5)$$

$$\boldsymbol{\lambda}_R^s = L_M(\mathbf{1}_s^s + \mathbf{1}_R^s) + L_L \mathbf{1}_R^s \qquad (2.6)$$

where

$$R_R = \gamma^2 R_r \qquad (2.7)$$

$$L_M = \gamma L_m = L_s \qquad (2.8)$$

$$L_L = \gamma L_{1s} + \gamma^2 L_{1r}. \qquad (2.9)$$

Eqs. (2.4) through (2.6), along with Eq. (1.24), describe the dynamic Γ equivalent circuit of an induction motor, shown in Figure 2.1.

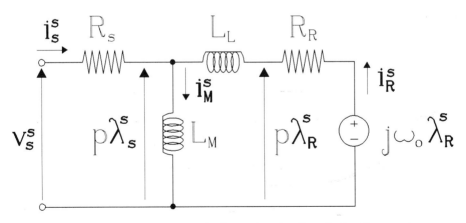

**Figure 2.1.** Dynamic Γ equivalent circuit of an induction motor.

The matrix voltage equation for the Γ circuit is

$$\begin{bmatrix} \mathbf{v}_s^s \\ 0 \end{bmatrix} = \begin{bmatrix} R_s + pL_M & pL_M \\ (p - j\omega_o)L_M & R_R + (p - j\omega_o)L_R \end{bmatrix} \begin{bmatrix} \mathbf{i}_s^s \\ \mathbf{i}_R^s \end{bmatrix} \quad (2.10)$$

where

$$L_R = L_L + L_M \quad (2.11)$$

and the torque equation is

$$T = \frac{P}{3} L_M Im(\mathbf{i}_s^s \mathbf{i}_R^{s*}) . \quad (2.12)$$

The corresponding, steady-state, per phase $\Gamma$ equivalent circuit is shown in Figure 2.2.

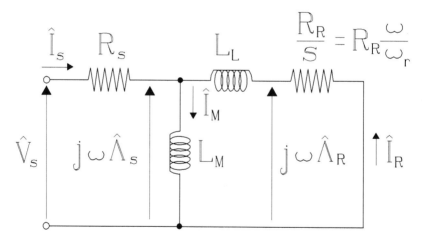

**Figure 2.2.** Steady-state $\Gamma$ equivalent circuit of an induction motor.

The voltage and torque equations for this circuit are:

$$\begin{bmatrix} \hat{V}_s \\ 0 \end{bmatrix} = \begin{bmatrix} R_s + j\omega L_M & j\omega L_M \\ j\omega L_M & \frac{R_R}{s} + j\omega L_R \end{bmatrix} \begin{bmatrix} \hat{I}_s \\ \hat{I}_R \end{bmatrix} \quad (2.13)$$

and

# SCALAR CONTROL OF INDUCTION MOTORS

$$T = 1.5 P L_M \text{Im}(\hat{I}_s \hat{I}_R^*). \tag{2.14}$$

Parameters of all the four elements of the Γ equivalent circuit can be determined from the no-load and blocked-rotor tests, supplemented by direct measurement of the stator resistance $R_s$. Here, this circuit will be used primarily to facilitate the analysis of CVH control.

## 2.2 Principles of the Constant Volts/Hertz Control

Under CVH control, stator voltage waveforms are assumed to be sinusoidal, but with adjustable frequency and magnitude. An analysis of motor operation in the steady state can thus be performed using the steady-state Γ equivalent circuit.

The same approach as in Example 1.1 can be used to determine the electrical power input to the rotor, $P_{elec}$, the mechanical power developed in the rotor, $P_{mech}$, and the developed torque, $T$. Specifically,

$$P_{elec} = 3 I_R^2 R_R \frac{\omega}{\omega_r} \tag{2.15}$$

$$P_{mech} = P_{elec} - 3 I_R^2 R_R \tag{2.16}$$

and

$$T = \frac{P_{mech}}{\omega_M}. \tag{2.17}$$

Current $I_R$ can be found from the Γ equivalent circuit as

$$I_R = |\hat{I}_R| = \left| \frac{-j\omega \Lambda_s}{R_R \frac{\omega}{\omega_r} + j\omega L_L} \right|$$

$$= \left| \frac{-j\Lambda_s}{\frac{R_R}{\omega_r} + j L_L} \right| = \frac{\Lambda_s}{\sqrt{\left(\frac{R_R}{\omega_r}\right)^2 + L_L^2}} \tag{2.18}$$

so

$$I_R^2 = \frac{\Lambda_s^2}{(\frac{R_R}{\omega_r})^2 + L_L^2} = \frac{\Lambda_s^2}{R_R^2} \frac{\omega_r^2}{(\tau\omega_r)^2+1} \quad (2.19)$$

where

$$\tau = \frac{L_L}{R_R}. \quad (2.20)$$

Substituting Eq. (2.19) in Eq. (2.15) gives

$$P_{elec} = 3\frac{\Lambda_s^2}{R_R} \frac{\omega\omega_r}{(\tau\omega_r)^2+1} \quad (2.21)$$

and

$$P_{mech} = 3\frac{\Lambda_s^2}{R_R} \frac{\omega_r(\omega-\omega_r)}{(\tau\omega_r)^2+1}. \quad (2.22)$$

As

$$\omega_M = \frac{2}{P}\omega_o = \frac{2}{P}(\omega-\omega_r) \quad (2.23)$$

then

$$T = 1.5P\frac{\Lambda_s^2}{R_R} \frac{\omega_r}{(\tau\omega_r)^2+1}. \quad (2.24)$$

The $T(\omega_r)$ curve described by Eq. (2.24) is shown in Figure 2.3 for an unlimited range of slip speed $\omega_r$. Analysing Eqs. (2.21) through (2.24), the following deductions can be made:

(1) If $\omega_r < 0$ ($s < 0$), then $\omega_o > \omega$, $T < 0$, $P_{mech} < 0$, and $P_{elec} < 0$, i.e., the developed torque opposes the motion of the rotor, mechanical power is consumed, and electrical power is delivered by the machine. This means that the machine operates as a generator, its rotor having been

forced to rotate with a supersynchronous speed.

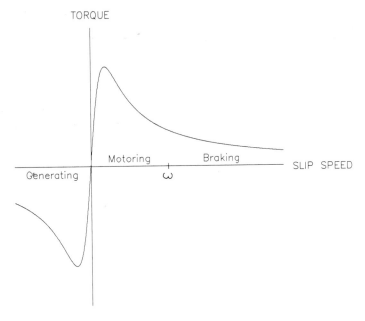

**Figure 2.3.** Torque versus slip speed characteristic of an induction motor with constant stator flux.

(2) If $\omega_r = 0$ ($s = 0$), then $\omega_o = \omega$, $T = 0$, $P_{mech} = 0$, and $P_{elec} = 0$, i.e., the motor operates on no load, at the synchronous speed given, in revolutions per minute (rpm), by

$$n_{M,syn} = 120 \frac{f}{P}. \qquad (2.25)$$

(3) If $0 < \omega_r < \omega$ ($0 < s < 1$), then $0 < \omega_o < \omega$, $T > 0$, $P_{mech} > 0$, and $P_{elec} > 0$, i.e., the machine operates as a motor.

(4) If $\omega_r = \omega$ ($s = 1$), then $\omega_o = 0$, $T = 0$, $P_{mech} = 0$, and $P_{elec} > 0$, i.e., the rotor is stalled and the developed, starting torque is

$$T_{start} = 1.5 P \frac{\Lambda_s^2}{R_R} \frac{\omega}{(\tau \omega)^2 + 1}. \qquad (2.26)$$

(5) If $\omega_r > \omega$ ($s > 1$), then $\omega_o < 0$, $T > 0$, $P_{mech} < 0$, and $P_{elec} > 0$, i.e., the

rotor is forced to rotate in the negative direction (against the rotating magnetic field of the stator) with the developed torque opposing the rotor motion, and both the mechanical and electrical power are consumed by the machine. This implies that the machine acts as a brake, dissipating both the mechanical and electrical power supplied to the rotor.

It must be pointed out that since the analysed machine is thought of as a motor, the mechanical power, $P_{mech}$, is assumed positive when the machine delivers this power, while the electrical power, $P_{elec}$, is assumed positive when the machine consumes this power.

Peak values of the $T(\omega_r)$ curve are

$$T_{peak} = \pm 0.75 P \frac{\Lambda_s^2}{L_L} \qquad (2.27)$$

which occur at

$$\omega_r(T_{peak}) = \pm \frac{1}{\tau}. \qquad (2.28)$$

The peak torque is often referred to as a breakdown, or pull-out, torque, because if it is exceeded by the load torque the motor stalls.

It is important to note that the slip speed, $\omega_r$, applies to a 2-pole motor. A $P$-pole motor can be thought of as a 2-pole motor with an attached gearbox having a gear ratio of $2/P$. If the synchronous speed of a $P$-pole motor is denoted by $\omega_{syn}$ and the slip speed of this motor by $\omega_{sl}$, then $\omega_{syn} = 2/P\,\omega$, $\omega_{sl} = 2/P\,\omega_r$, and $\omega_M = 2/P\,\omega_o$.

When the magnitude, $V_s$, of the stator voltage is controlled in such a way that the magnitude, $\Lambda_s$, of the stator flux is maintained constant, independently of changes in the supply frequency, $f$, then the torque developed in the motor depends on the slip speed, $\omega_r$, only. At the same time, it follows from Eq. (2.23) that the motor speed is a linear function of the frequency, as given by

$$\omega_M = \frac{2}{P}(2\pi f - \omega_r) \qquad (2.29)$$

or

$$n_M = \frac{60}{2\pi}\omega_M = \frac{120}{P}\left(f - \frac{\omega_r}{2\pi}\right). \qquad (2.30)$$

If the voltage drop across $R_s$ (see Figure 2.2) is negligible in comparison with $V_s$, then

$$\hat{V}_s \approx j\omega\hat{\Lambda}_s \qquad (2.31)$$

i.e.,

$$V_s \approx \omega\Lambda_s \qquad (2.32)$$

from which

$$\Lambda_s \approx \frac{V_s}{\omega} = \frac{1}{2\pi}\frac{V_s}{f}. \qquad (2.33)$$

Eq. (2.33) provides an explanation of the term "Constant Volts/Hertz Control". If the ratio $V_s/f$ remains constant with the changes of $f$, then $\Lambda_s$ remains constant too and the torque is independent of the supply frequency. In practice, the stator voltage to frequency ratio is usually based on the rated values of these variables. However, at low levels of supply frequency, when the stator voltage also becomes low, the voltage drop across the stator resistance cannot be neglected and must be compensated by increasing $V_s$ above the value resulting from the CVH condition, $V_s/f =$ *const*.

At frequencies higher than the rated value, the CHV condition also cannot be satisfied since, in order to avoid insulation breakdown, the stator voltage may not exceed its rated value. Therefore, in this range of frequencies, the stator voltage is maintained at the rated level, and the stator flux decreases with increases in frequency. Unavoidably, the developed torque decreases correspondingly.

The variation of voltage in this type of control is illustrated in Figure 2.4 showing the constant voltage to frequency line (dashed) and the actual $V_s$ versus $f$ curve (continuous) to be followed as the supply frequency is varied.

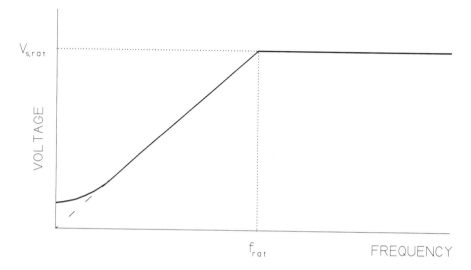

**Figure 2.4.** Voltage versus frequency curve under the CVH control.

## 2.3 Scalar Speed Control System

The general configuration of a speed control system based on the previously described scalar, CVH control method is shown in Figure 2.5. The motor is operated from a voltage-controlled inverter which constitutes a source of a three-phase a.c. voltage with adjustable frequency and magnitude. The inverter is essentially a network of semiconductor power switches and diodes supplied from a d.c. source, which in practice is a rectifier with a low-pass power filter (d.c. link). The switches allow to alternate positive and negative potentials of the d.c. source to be applied to the stator terminals of the motor in such way that the desired fundamental components of the phase voltages, $v_{as}$, $v_{bs}$, and $v_{cs}$, are produced. The diodes provide additional paths for the a.c. currents supplied to the stator when they cannot flow through the switches that are not conducting. See Chapter 5 for more details on inverter operation.

The control system includes a feedback loop from the actual motor speed signal, $\omega_M$, obtained from a speed sensor on the rotor shaft. This signal is compared with the reference speed signal, $\omega_M^*$, and the resulting speed error, $\Delta\omega_M$, is applied to a slip controller whose output represents a reference slip speed signal, $\omega_{sl}^*$. The static characteristic of the controller limits this speed to a value somewhat less than that corresponding to the peak torque of the motor. In this way, the motor is forced to operate within the area between the peaks of the developed torque, i.e., on the steep portion of the torque curve (see Figure 2.3). This portion is often called "stable", since in an uncontrolled drive system, a load change results in a speed change such that the corresponding motor torque matches the new load torque, and the system settles down to a new stable operating point. Setting the limit on the slip speed in the vicinity of the peak torque point, provides fast response of the drive system to changes in the reference speed.

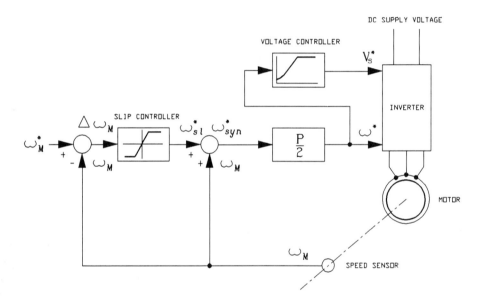

**Figure 2.5.** General configuration of a scalar speed control system.

Signals $\omega_{sl}^*$ and $\omega_M$ are added together to produce a reference synchronous speed signal, $\omega_{syn}^*$, proportional, with the coefficient of $P/2$, to the required supply radian frequency, $\omega^*$. A voltage controller generates an appropriate value of the reference magnitude signal, $V^*$, of the stator voltage to be supplied by the inverter (see Figure 2.4).

If the torque is plotted as a function of motor speed, a mechanical characteristic of the motor is obtained. A family of such characteristics at various values of the supply frequency is shown in Figure 2.6. As mentioned previously, the peak torque remains constant at frequencies below rated frequency, and decreases with increase in frequency above the rated value, where the CHV condition is no longer satisfied.

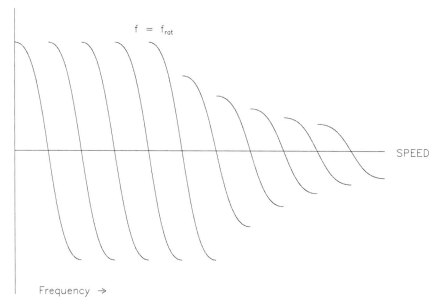

**Figure 2.6.** Mechanical characteristics of an induction motor with CVH control.

## 2.4 The Γ' Equivalent Circuit of an Induction Motor

To explain the scalar torque control (TC) system, it is useful to introduce yet another equivalent circuit of the motor, this called inverse-Γ, or Γ', circuit. In a similar manner to the T→Γ circuit transformation (see Section 2.1), the T→Γ' transformation is performed by introducing a transformation coefficient γ' defined as

$$\gamma' \equiv \frac{L_m}{L_r}. \qquad (2.34)$$

Then, following an analogous procedure as that in Section 2.1, a dynamic equivalent citcuit of an induction motor, shown in Figure 2.7, is obtained, in which

$$R'_R = \gamma'^2 R_r \qquad (2.35)$$

# SCALAR CONTROL OF INDUCTION MOTORS

$$L'_M = \gamma' L_m \qquad (2.36)$$

$$L'_L = L_{1s} + \gamma' L_{1r} \qquad (2.37)$$

and

$$\begin{bmatrix} \mathbf{i}_R^{s'} \\ \lambda_R^{s'} \end{bmatrix} = \begin{bmatrix} \gamma'^{-1} & \\ & \gamma' \end{bmatrix} \begin{bmatrix} \mathbf{i}_r^{s} \\ \lambda_r^{s} \end{bmatrix}. \qquad (2.38)$$

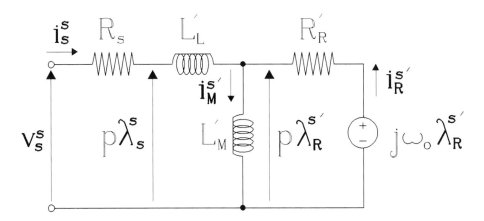

**Figure 2.7.** Dynamic Γ' equivalent circuit of an induction motor.

The matrix voltage equation for this circuit is

$$\begin{bmatrix} \mathbf{v}_s^{s} \\ 0 \end{bmatrix} = \begin{bmatrix} R_s + pL'_s & pL'_M \\ (p - j\omega_o) L'_M & R'_R + (p - j\omega_o) L'_M \end{bmatrix} \begin{bmatrix} \mathbf{i}_s^{s} \\ \mathbf{i}_R^{s'} \end{bmatrix} \qquad (2.39)$$

where

$$L'_s = L'_L + L'_M \qquad (2.40)$$

and the torque equation is

$$T = \frac{P}{3}L'_M \text{Im}(\mathbf{1_s^s 1_R^{s'*}}) . \quad (2.41)$$

The inverse-Γ equivalent circuit of an induction motor operating in the steady state is shown in Figure 2.8. The voltage and torque equations are

$$\begin{bmatrix} \hat{V}_s \\ 0 \end{bmatrix} = \begin{bmatrix} R_s + j\omega L'_s & j\omega L'_M \\ j\omega L'_M & \dfrac{R'_R}{s} + j\omega L'_M \end{bmatrix} \begin{bmatrix} \hat{I}_s \\ \hat{I}'_R \end{bmatrix} \quad (2.42)$$

and

$$T = 1.5 P L'_M \text{Im}(\hat{I}_s \hat{I}'^*_R) . \quad (2.43)$$

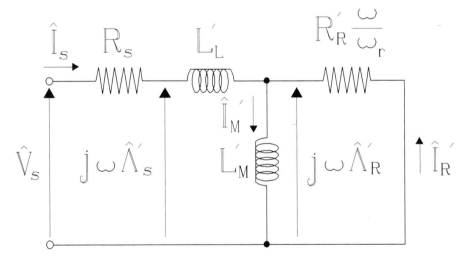

Figure 2.8. Steady-state Γ' equivalent circuit of an induction motor.

## 2.5 Principles of the Torque Control

In place of Eq. (2.43), another formula for the motor torque can be derived.

## SCALAR CONTROL OF INDUCTION MOTORS

Since

$$T = \frac{P}{2}\frac{P_{mech}}{\omega_o} \qquad (2.44)$$

where

$$\begin{aligned}P_{mech} &= 3I_R^{\prime 2}R_R^\prime\frac{\omega}{\omega_r} - 3I_R^{\prime 2}R_R^\prime \\ &= 3I_R^{\prime 2}R_R^\prime\frac{\omega-\omega_r}{\omega_r} = 3I_R^{\prime 2}R_R^\prime\frac{\omega_o}{\omega_r}\end{aligned} \qquad (2.45)$$

then

$$T = 1.5PI_R^{\prime 2}\frac{R_R^\prime}{\omega_r}. \qquad (2.46)$$

From the circuit,

$$I_R^\prime = I_M^\prime\frac{\omega L_M^\prime}{R_R^\prime\frac{\omega}{\omega_r}} = I_M^\prime\frac{\omega_r L_M^\prime}{R_R^\prime}. \qquad (2.47)$$

hence, the developed torque can be expressed as

$$\begin{aligned}T &= 1.5PI_R^\prime\frac{R_R^\prime}{\omega_r}\times I_R^\prime \\ &= 1.5PI_R^\prime\frac{R_R^\prime}{\omega_r}\times I_M^\prime\frac{\omega_r L_M^\prime}{R_R^\prime} \\ &= 1.5PL_M^\prime I_R^\prime I_M^\prime.\end{aligned} \qquad (2.48)$$

If the magnitude (r.m.s.) of the magnetizing current, $I_M^\prime$, is maintained constant, then the torque is directly proportional to the magnitude, $I_R^\prime$, of the rotor current. On the other hand, the magnitude, $\Lambda_R^\prime$, of the rotor flux is directly proportional to $I_M^\prime$. In effect, the stator current, $\hat{I}_s = \hat{I}_M^\prime - \hat{I}_R^\prime$, can be thought of as a sum of the so-called flux-producing current, $\hat{I}_{s\Phi} = \hat{I}_M^\prime$, and torque-producing current, $\hat{I}_{sT} = -\hat{I}_R^\prime$. When the both components of the stator current are independently adjusted, then, at a constant $I_{s\Phi}$, the motor becomes a linear $I_{sT} \rightarrow T$ converter, as illustrated in Figure 2.9.

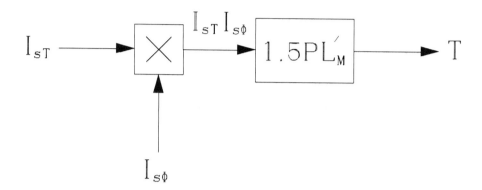

**Figure 2.9.** Block diagram of a torque-controlled induction motor in the steady state.

To synthesize a stator current such that its flux-producing and torque-producing components equal certain desired (reference) values $I_{s\Phi}^*$ and $I_{sT}^*$, respectively, a current controller is needed. The reference magnitude, $I_s^*$, and radian frequency, $\omega^*$, of the stator current have to be supplied to this controller. The reference magnitude is

$$I_s^* = \sqrt{I_{s\Phi}^{*2} + I_{sT}^{*2}} \qquad (2.49)$$

while to determine $\omega^*$, Eq. (2.47) can be rewritten as

$$I_{sT}^* = I_{s\Phi}^* \frac{\omega_r^* L_M'}{R_R'} \qquad (2.50)$$

and rearranged, to give

$$\omega_r^* = \frac{R_R'}{L_M'} \frac{I_{sT}^*}{I_{s\Phi}^*}. \qquad (2.51)$$

From Eqs. (2.34) through (2.37),

SCALAR CONTROL OF INDUCTION MOTORS 59

$$\frac{R'_R}{L'_M} = \frac{\gamma'^2 R_r}{\gamma' L_m} = \frac{L_m}{L_r}\frac{R_r}{L_m} = \frac{R_r}{L_r} = \frac{1}{\tau_r} \quad (2.52)$$

where $\tau_r$ denotes the time constant of the rotor. Hence, the reference slip speed of an equivalent 2-pole motor is

$$\omega^*_r = \frac{1}{\tau_r}\frac{I^*_{sT}}{I^*_{s\Phi}} \quad (2.53)$$

while the reference frequency of the stator current of a $P$-pole motor is

$$\omega^* = \omega_o + \omega^*_r = \frac{P}{2}\omega_M + \frac{1}{\tau_r}\frac{I^*_{sT}}{I^*_{s\Phi}}. \quad (2.54)$$

## 2.6 Scalar Torque Control System

A scalar torque control system, based on the principles explained in Section 2.5, is shown in Figure 2.10. Calculator 1 performs the computation of the reference values $I^*_{sT}$ and $I^*_{s\Phi}$ corresponding to the reference torque and rotor flux signals $T^*$ and $\Lambda'^*_R$, respectively. The reference torque-producing current is calculated as

$$I^*_{sT} = \frac{T^*}{1.5 P \Lambda'^*_R} \quad (2.55)$$

and the reference flux-producing current as

$$I^*_{s\Phi} = \frac{\Lambda'^*_R}{L'_M}. \quad (2.56)$$

Using Eqs. (2.49) and (2.54), calculator 2 determines the reference values of the magnitude, $I^*_s$, and of the radian frequency, $\omega^*$, of the stator current, based on the $I^*_{sT}$, $I^*_{s\Phi}$, and $\omega_M$ signals, the last supplied by a sensor measuring the speed of the motor. In practical systems, a single microprocessor performs the functions of both the calculators shown, for clarity, as separate blocks in the diagram in Figure 2.10.

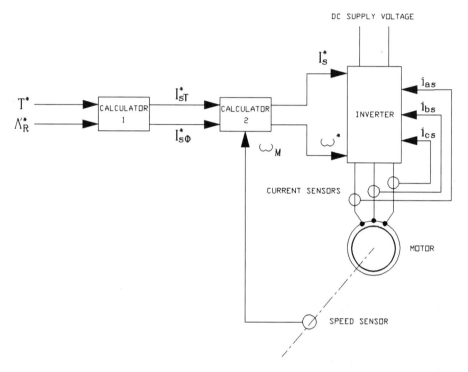

**Figure 2.10.** Block diagram of a scalar torque control system.

The inverter feeding the motor must be of the current-controlled type, which means that there must be feedback from the stator phase currents, $i_{as}$, $i_{bs}$, and $i_{cs}$, provided by current sensors installed in the supply leads. This allows continuous control of these currents in accordance with the reference signals $I_s^*$ and $\omega^*$.

The torque developed by an induction motor is limited by the maximum allowable values of the stator voltage and current. A voltage in excess of the rated value can cause an insulation breakdown while an excessive current would, over a period of time, result in overheating and possible damage to the windings. Only under transient conditions and for a limited time, can a higher than rated current (but not voltage) and, consequently, a higher than rated torque, be permitted.

The allowable values of the rotor flux, $\Lambda_R'$, and torque, $T$, at various speeds, $\omega_M$, of a torque-controlled motor represent the so-called safe operating area (SOAR) of the motor. Under given operating conditions, the

## SCALAR CONTROL OF INDUCTION MOTORS

r.m.s. value of the stator currents may not exceed a certain limit, which depends on these conditions. For instance, if the motor operates in a continuous mode, the currents should not be higher than the rated value, while in an intermittent mode of operation, short-term overloads, up to a specified allowable value, $I_{s,all}$, are permitted. At the same time, the voltages suplied to the stator windings may never exceed the rated value, $V_{s,rat}$.

Using Eqs. (2.49), (2.55), and (2.56), and assuming that the actual motor variables are equal to the corresponding reference values, the current condition, $I_s \leq I_{s,all}$, can be expressed as

$$\sqrt{\left(\frac{T}{1.5 P \Lambda_R'}\right)^2 + \left(\frac{\Lambda_R'}{L_M'}\right)^2} \leq I_{s,all} \qquad (2.57)$$

from which,

$$T \leq 1.5 P \Lambda_R' \sqrt{I_{s,all}^2 - \left(\frac{\Lambda_R'}{L_M'}\right)^2}. \qquad (2.58)$$

To express the voltage condition, $V_s \leq V_{s,rat}$ in terms of the torque, $T$, and flux, $\Lambda_R'$, the steady-state Γ' equivalent circuit of the motor can be used. Since, as seen in Figure 2.8,

$$\hat{V}_s = (R_s + j\omega L_L') \hat{I}_s + j\omega \hat{\Lambda}_R' \qquad (2.59)$$

where

$$\hat{I}_s = I_{sT} - jI_{s\phi} \qquad (2.60)$$

and

$$\hat{\Lambda}_R' = -j\Lambda_R' \qquad (2.61)$$

then

$$\hat{V}_s = R_s I_{sT} + \omega (L_L' I_{s\phi} + \Lambda_R') \\ + j(\omega L_L' I_{sT} - R_s I_{s\phi}). \quad (2.62)$$

Substituting Eqs. (2.55) and (2.56) in Eq. (2.62), and determining the absolute value (magnitude), $V_s$, of phasor $\hat{V}_s$, the voltage condition can be expressed in a form of the quadratic inequality

$$\frac{R_s^2 + (\omega L_L')^2}{2.25 P^2} \left(\frac{T}{\Lambda_R'}\right)^2 + \frac{R_s \omega}{0.75 P} T \\ + \left\{R_s^2 + [\omega (L_L' + L_M')]^2\right\} \left(\frac{\Lambda_R'}{L_M'}\right)^2 - V_{s,rat}^2 \leq 0. \quad (2.63)$$

Eqs. (2.58) and (2.63) express the SOAR conditions. Since three variables, i.e., $T$, $\Lambda_R'$, and $\omega$, are involved, the SOAR can only be illustrated by a three-dimensional graph. Such a graph is shown in Figure 2.11, for positive values of the torque, $T$, and rotor flux, $\Lambda_R'$, and $I_{s,all} = I_{s,rat}$. The SOAR is represented there by the space below the grid surface which constitutes a locus of the maximum allowable values of the torque. The "precipice" seen at the left-hand side of the graph results from the voltage condition, since, at a given frequency, certain values of the rotor flux are not available because of the limited magnitude of the stator voltage.

In practice, the rotor flux is usually controlled in a similar manner as the stator flux in the speed control systems. As illustrated in Figure 2.12, above the rated frequency, $\omega_{rat}$, the flux is adjusted in inverse proportion to frequency, i.e.,

$$\Lambda_R'(\omega) = \begin{cases} \Lambda_R'(\omega_{rat}) & \text{for } \omega \leq \omega_{rat} \\ \dfrac{\omega_{rat}}{\omega} \Lambda_R'(\omega_{rat}) & \text{for } \omega > \omega_{rat}. \end{cases} \quad (2.64)$$

The SOAR graph for such a control mode and $I_{s,all} = I_{s,rat}$ is shown in Figure 2.13. The flux axis represents values of the rotor flux, $\Lambda_R'(\omega_{rat})$, at the rated frequency. It can be seen that the precipice of Figure 2.11 has disappeared and, above the rated frequency, the SOAR is now limited by a gently sloped hyperbolical surface. In a similar manner to that for a speed-controlled motor (see Figure 2.6), the available torque decreases with the

SCALAR CONTROL OF INDUCTION MOTORS 63

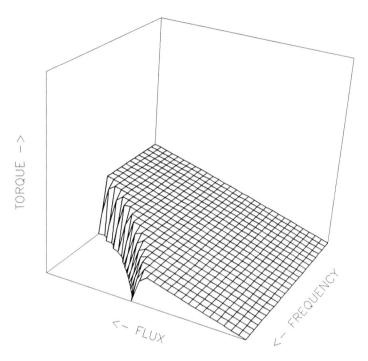

**Figure 2.11.** Three-dimensional graph of the safe operating area for a torque-controlled induction motor with the frequency-independent flux control: $I_{s,all} = I_{s,rat}$.

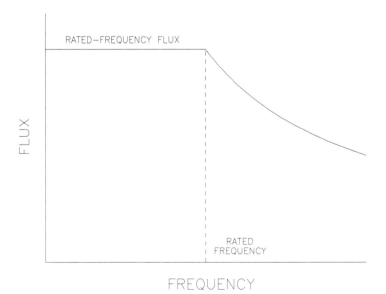

**Figure 2.12.** Rotor flux versus frequency curve under the torque control.

increase in the motor speed above the rated value (field-weakening area). This is illustrated in Figure 2.14 for $\Lambda_R'(\omega_{rat}) = \Lambda_{R,rat}'$. Readers familiar with operating characteristics of d.c. motors will notice the close analogy between the torque versus frequency characteristics of the torque-controlled induction motor and the torque versus speed characteristics of shunt or separately excited d.c. machines.

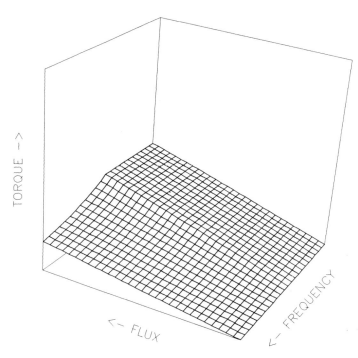

**Figure 2.13.** Three-dimensional graph of the safe operating area for a torque-controlled induction motor with the frequency-dependent flux control: $I_{s,all} = I_{s,rat}$.

If the maximum allowable stator current is significantly higher than the rated current, which is permitted for short overload periods, the SOAR becomes somewhat more complicated, as shown in Figure 2.15 for $I_{s,all} = 3I_{s,rat}$.

Three-dimensional SOAR graphs are inconvenient to use. Therefore, two-dimensional SOAR curves can be employed instead, to represent the relation between the maximum allowable torque and rotor flux, at fixed values of the frequency. Such curve, calculated for the motor used in the examples and simulations, is shown in Figure 2.16 for $I_{s,all} = 3I_{s,rat}$. The lower the frequency, the higher torque is available from the motor. For example, a torque of up to 190 N m, at $\Lambda_R' = 0.3$ Wb, is feasible, if the

# SCALAR CONTROL OF INDUCTION MOTORS

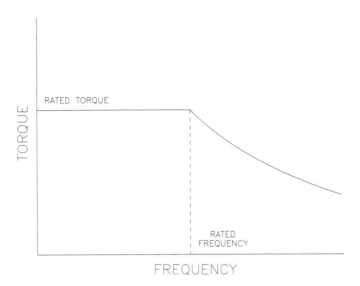

**Figure 2.14.** Torque versus frequency characeristic of a torque-controlled induction motor with the frequency-dependent flux control.

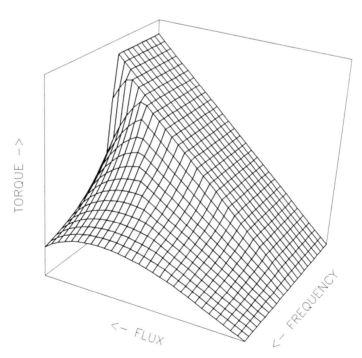

**Figure 2.15.** Three-dimensional graph of the safe operating area for a torque-controlled induction motor: frequency-dependent flux control: $I_{s,all} = 3I_{s,rat}$.

not exceed 50 Hz. At 60 Hz, this maximum allowable value of the torque drops to 150 N m, at $\Lambda_R' = 0.24$ Wb. Finally, at a frequency of 120 Hz, which is twice as high as the rated value, only 50 N m of the torque is available, which is less than the rated torque of 61.2 N m, even though a 200% current overload is permitted.

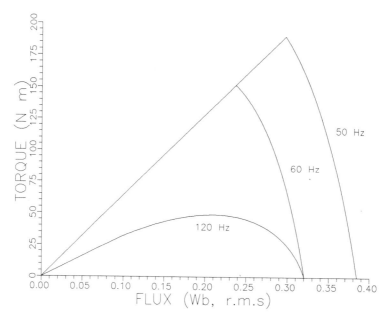

**Figure 2.16.** Safe operating area of the example motor: $I_{s,all} = 3I_{s,rat}$.

## 2.7 Examples and Simulations

The following examples and simulations are included in order to clarify the concepts of the $\Gamma$ and $\Gamma'$ equivalent circuits of an induction motor, and to illustrate the operation of a motor in scalar speed and torque control systems.

### Example 2.1.

Determine the $\Gamma$ equivalent circuit of the example induction motor and calculate the rated torque developed by the motor. Compare the result with that obtained in Example 1.1.

*Solution*

The T→Γ transformation coefficient, $\gamma$, is given by

SCALAR CONTROL OF INDUCTION MOTORS

$$\gamma = \frac{L_s}{L_m} = \frac{0.0424}{0.041} = 1.034$$

and the leakage reactances of the motor are

$$L_{1s} = \frac{X_{1s}}{\omega} = \frac{0.524}{377} = 0.00139 \; H/ph$$

$$L_{1r} = \frac{X_{1r}}{\omega} = \frac{0.279}{377} = 0.00074 \; H/ph.$$

Hence,

$$R_R = \gamma^2 R_r = 1.034^2 \times 0.156 = 0.167 \; \Omega/ph$$

$$L_L = \gamma L_{1s} + \gamma^2 L_{1r} = 1.034 \times .00139 + 1.034^2 \times 0.00074$$
$$= 0.00223 \; H/ph$$

while

$$L_M = L_s = 0.0424 \; H/ph$$

and

$$L_R = L_L + L_M = 0.00223 + 0.0424 = 0.04463 \; H/ph.$$

The steady-state voltage equation (2.13) is

$$\begin{bmatrix} 127 \\ 0 \end{bmatrix} = \begin{bmatrix} 0.294 + j377 \times 0.0424 & j377 \times 0.0424 \\ j377 \times 0.0424 & \frac{0.167}{0.03} + j377 \times 0.04463 \end{bmatrix} \cdot \begin{bmatrix} \hat{I}_s \\ \hat{I}_R \end{bmatrix}$$

from which

$$\begin{bmatrix} \hat{I}_s \\ \hat{I}_R \end{bmatrix} = \begin{bmatrix} 0.294+j15.985 & j15.985 \\ j15.985 & 5.567+j16.826 \end{bmatrix}^{-1} \begin{bmatrix} 127 \\ 0 \end{bmatrix}$$

$$= \begin{bmatrix} 23.8\angle-25.5° \\ 21.5\angle172.8° \end{bmatrix} A/ph.$$

Finally,

$$T = 1.5 \times 6 \times 0.0424 \times Im(23.8\angle-25.5° \times 21.5\angle-172.8°)$$
$$= 195\sin(-198.3°) = \mathbf{61.2\ N\ m}.$$

The same result as in Example 1.1 has been obtained. Notice that the calculated current phasors also agree with those in Example 1.1, as

$$\hat{I}_r = \gamma \hat{I}_R = 1.034 \times 21.5 \angle 172.8°$$
$$= 22.2 \angle 172.8°\ A/ph.$$

**Example 2.2.**

At rated stator flux and frequency, determine the starting torque and compare it with the rated torque.

*Solution*

Firstly, the rated stator flux, $\Lambda_s$, and time constant $\tau$ are calculated as

$$\Lambda_s = |L_M \hat{I}_{M,rat}| = |L_M(\hat{I}_{s,rat} + \hat{I}_{R,rat})|$$
$$= |0.0424(23.8\angle-25.5° + 21.5\angle172.8°)|$$
$$= 0.32\ Wb/ph$$

$$\tau = \frac{L_L}{R_R} = \frac{0.00223}{0.167} = 0.0134\ sec.$$

Consequently, from Eq. (2.26),

# SCALAR CONTROL OF INDUCTION MOTORS

$$T_{start} = 1.5 \times 6 \times \frac{0.32^2}{0.167} \times \frac{377}{(0.0134 \times 377)^2 + 1} = \mathbf{78.4\ N}$$

The starting torque is higher than the rated torque, which is typical for practical induction motors. In this way, they are able of starting even on full load.

### Example 2.3.

Calculate the voltage versus frequency curve, corresponding to that shown in Figure 2.4, for the voltage controller.

*Solution*

From the Γ equivalent circuit,

$$\hat{V}_s = j\omega\hat{\Lambda}_s + R_s \hat{I}_s$$

where

$$\hat{I}_s = \hat{I}_M - \hat{I}_R.$$

Since

$$\hat{I}_M = \frac{j\omega\hat{\Lambda}_s}{j\omega L_M} = \frac{\hat{\Lambda}_s}{L_M}$$

and

$$\hat{I}_R = -\frac{j\omega\hat{\Lambda}_s}{j\omega L_L + R_R \frac{\omega}{\omega_r}} = -\frac{\hat{\Lambda}_s}{L_L - j\frac{R_R}{\omega_r}}$$

then

$$\hat{V}_s = \hat{\Lambda}_s [j\omega + R_s(\frac{1}{L_M} + \frac{1}{L_L - j\frac{R_R}{\omega_r}})]$$

i.e.,

$$V_s = \Lambda_s | j\omega + R_s ( \frac{1}{L_M} + \frac{1}{L_L - j\frac{R_R}{\omega_r}} ) | . \quad (2.65)$$

Substituting the rated stator flux $\Lambda_s$ of 0.32 Wb/ph and the rated slip speed $\omega_r$ of 377 - 365.7 = 11.3 rad/sec (see Example 1.2) in Eq. (2.62), the equation of the curve is obtained as

$$V_s = 0.32 | j 2\pi f + 0.294 ( \frac{1}{0.0424} + \frac{1}{0.00223 - \frac{j 0.167}{11.3}} ) |$$

$$= 0.32 \sqrt{(2\pi f + 19.45)^2 + 97.36} \ \ V/ph$$

for the frequency range 0 to 60 Hz. Above 60 Hz, the stator voltage must be maintained at the rated level of 127 V/ph. The resultant voltage control curve is shown in Figure 2.17.

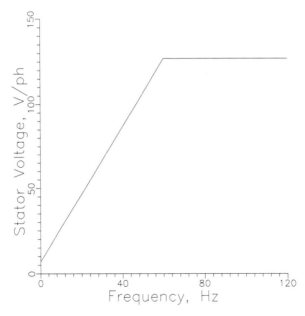

**Figure 2.17.** Voltage control curve for the example motor.

SCALAR CONTROL OF INDUCTION MOTORS 71

### Example 2.4.

At rated supply voltage and frequency, and within the normal operating range of a scalar-controlled induction motor, i.e., $-1/\tau < \omega_r < 1/\tau$, determine and plot the following quantities of the example motor versus the speed of the motor:

(a) input power, $P_{inp}(n_M)$,
(b) output power, $P_{out}(n_M)$,
(c) torque, $T(n_M)$,
(d) stator current, $I_s(n_M)$,
(e) efficiency, $\eta(n_M)$,
(f) power factor, $PF(n_M)$.

*Solution*

The required characteristics are determined by solving Eqs. (2.13) (substituting $\omega_r/\omega$ for $s$) and (2.14) for consecutive values of $\omega_r$ within the range $-1/\tau$ to $1/\tau$. The motor speed versus slip speed relation is then given by Eq. (2.30). If $\omega_r < 0$, i.e., the machine operates as a generator,

$$P_{inp} = -T\omega_M \qquad (2.66)$$

and

$$P_{out} = -3Re(\hat{V}_s \hat{I}_s^*). \qquad (2.67)$$

Conversely, if the machine operates as a motor, i.e., $0 < \omega_r < \omega$, then

$$P_{inp} = 3Re(\hat{V}_s \hat{I}_s^*) \qquad (2.68)$$

and

$$P_{out} = T\omega_M. \qquad (2.69)$$

The efficiency is calculated as

$$\eta = \frac{P_{out}}{P_{inp}} \qquad (2.70)$$

and the power factor is

$$PF = \frac{P_{out}}{3V_s I_s} \quad (2.71)$$

for $\omega_r < 0$, and

$$PF = \frac{P_{inp}}{3V_s I_s} \quad (2.72)$$

for $0 < \omega_r < \omega$.

The input and output powers are shown in Figure 2.18. Since in the employed model of an induction motor the rotational losses are not accounted for, the input power at the synchronous speed of 1200 rpm represents only the power loss in the stator resistance, in this case equal to about 50 W and not discernible in the plot.

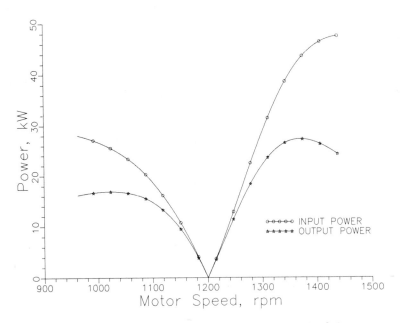

**Figure 2.18.** Input and output power versus speed of the motor.

The mechanical characteristic of the motor, i.e., developed torque

SCALAR CONTROL OF INDUCTION MOTORS 73

versus speed, is shown in Figure 2.19, while the stator current versus speed curve is shown in Figure 2.20. High absolute values of slip speed are acompanied by high values of stator current. Therefore, at the direct-on-line starting, as illustated in Simulation 1.1, the stator current tends to be many times higher than the rated current. As seen in the subsequent simulations, this problem can be alleviated by scalar-controlled, adjustable-frequency starting.

The efficiency and power factor characteristics are shown in Figure 2.21. It can be observed that the motor has been designed in such way that in the vicinity of the rated speed of the motor both the last performance indicators have the desired high values.

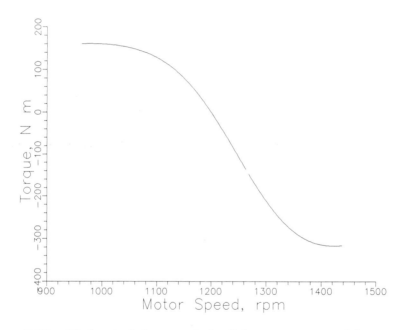

**Figure 2.19.** Mechanical characteristic of the motor at rated frequency.

### Example 2.5.

Determine the Γ' equivalent circuit of the example motor and calculate the developed torque, $T$, and rotor flux, $\Lambda_R'$, under rated operating conditions.

*Solution*

The transformation coefficient, $\gamma$, ... given by

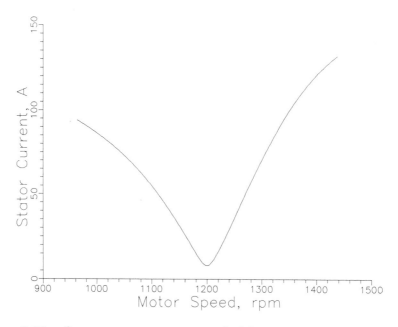

**Figure 2.20.** Stator current versus speed of the motor at rated frequency.

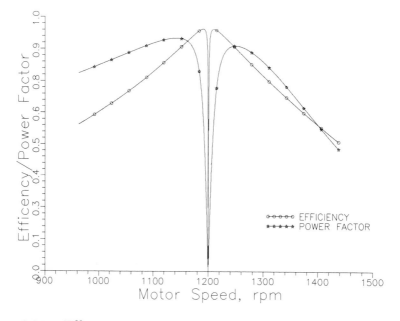

**Figure 2.21.** Efficiency and power factor versus speed of the motor at rated frequency.

$$\gamma' = \frac{L_m}{L_r} = \frac{0.041}{0.0417} = 0.9832.$$

Hence,

$$R'_R = \gamma'^2 R_r = 0.9832^2 \times 0.156 = 0.151 \ \Omega/ph$$

$$L'_M = \gamma' L_m = 0.9832 \times 0.041 = 0.0403 \ H/ph$$

$$L'_L = L_{ls} + \gamma' L_{lr} = 0.00139 + 0.9832 \times 0.00074$$
$$= 0.00212 \ H/ph$$

$$L'_s = L'_L + L'_M = 0.00212 + 0.0403 = 0.04242 \ H/ph.$$

Consequently, the matrix voltage equation (2.42) is

$$\begin{bmatrix} 127 \\ 0 \end{bmatrix} = \begin{bmatrix} 0.294 + j15.996 & j15.193 \\ j15.193 & 5.033 + j15.193 \end{bmatrix} \begin{bmatrix} \hat{I}_s \\ \hat{I}'_R \end{bmatrix}$$

which yields

$$\begin{bmatrix} \hat{I}_s \\ \hat{I}'_R \end{bmatrix} = \begin{bmatrix} 23.8\angle -25.8° \\ 22.6\angle 172.8° \end{bmatrix} A/ph.$$

The developed torque can be calculated from Eq. (2.43) as

$$T = 1.5 \times 6 \times 0.0403 \times 22.6 \times 7.47 = \mathbf{61.2 \ N \ m}$$

or from Eq. (2.48), in which case the magnitude, $I'_M$, of the magnetizing current must be determined first. Since

$$I'_M = |\hat{I}'_s + \hat{I}'_R| = |23.8\angle-25.8°+22.6\angle172.8°|$$
$$= 7.47 \text{ A/ph}$$

then the torque is

$$T = 1.5\times6\times0.0403\times Im(23.8\angle-25.8°\times22.6\angle-172.8°)$$
$$= \mathbf{61.2\ N\ m}$$

and the rotor flux is

$$\Lambda'_R = L'_M I'_M = 0.0403\times7.47 = \mathbf{0.3\ Wb/ph}.$$

### Example 2.6.

Calculate the magnitude and frequency of the stator current such that the motor operates with the rated torque of 61.2 N m, at the rated speed of 1164 rpm, and the rated rotor flux determined in Example 2.5.

### *Solution*

Based on Eqs. (2.55) and (2.56), the torque-producing and flux-producing currents are

$$I^*_{sT} = \frac{61.2}{1.5\times6\times0.3} = 22.6 \text{ A/ph}$$

$$I^*_{s\Phi} = \frac{0.3}{0.0403} = 7.47 \text{ A/ph}.$$

Hence,

$$I^*_s = \sqrt{22.6^2+7.47^2} = \mathbf{23.8\ A/ph}.$$

The rotor time constant is

$$\tau_r = \frac{L_r}{R_r} = \frac{0.0417}{0.156} = 0.267 \text{ sec}$$

and the rated angular speed of the rotor is

$$\omega_{M,rat} = \frac{2\pi}{60} \times 1164 = 121.9 \ rad/sec.$$

Consequently, from Eq. (2.54),

$$\omega^* = \frac{6}{2} \times 121.9 + \frac{1}{0.267} \times \frac{22.6}{7.4} = \mathbf{377 \ rad/sec}$$

i.e., $f^* = 60$ Hz.

Unsurprisingly, the obtained results represent the rated magnitude and frequency of the stator current.

### Simulation 2.1. Speed-Controlled Starting

To show the difference between the dynamic behavior of a speed-controlled induction motor and that of an uncontrolled motor, the starting performance of the example motor, when controlled in the system shown in Figure 2.5, is simulated. Load parameters are the same as in Simulation 1.1 illustrating direct-on-line starting, and the reference speed is 1183 rpm (steady-state speed in Simulation 1.1). The slip speed limit of the slip controller is set at 67.1 rad/sec. No limits on current overload are assumed. The magnitude of the stator voltage is adjusted in accordance to the equation derived in Example 2.3. The sampling frequency of the digital speed control system is 10 kHz, i.e., adjustments to the magnitude and frequency of the stator voltage are made every 0.1 msec. This condition is simulated by setting the time step of the simulation program to that value. The inverter is assumed to produce ideal a.c. currents in the stator.

The torque and speed of the motor are shown in Figure 2.22. Comparing these with those corresponding to direct-on-line, fixed-frequency starting (Figure 1.15), a substantial improvement in the starting conditions can be noted. Torque oscillations are significantly less and the starting time is reduced by half. However, the reference speed is not reached, as the motor speed stabilizes itself at the 1172 rpm level because of the inherent steady-state control error in the speed control loop. To avoid this error, a proportional-plus-integral (PI) controller is usually incorporated before the slip controller.

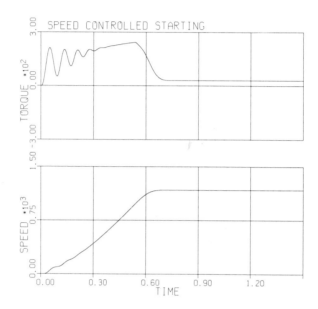

**Figure 2.22.** Torque and speed of the motor with speed-controlled starting.

The stator and rotor current waveforms are shown in Figure 2.23. A gradual increase of the supply frequency during the acceleration of the motor is easily observed.

### Simulation 2.2. Scalar Speed Control - Load Changes

To illustrate the impact of a PI speed controller, mentioned in Simulation 2.1, on the operation of the drive system, a scalar speed control system equipped with such a regulator is simulated under the same rapid load changes as those in Simulation 1.2. Figure 2.24 shows the torque and speed of the motor, while the stator and rotor currents are depicted in Figure 2.25. It can be seen that the load changes hardly affect the speed of the motor at all. Note that different torque and current transients to those in Simulation 1.2 are now present.

SCALAR CONTROL OF INDUCTION MOTORS 79

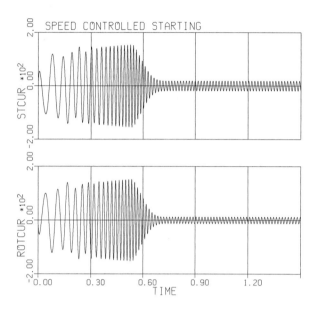

**Figure 2.23.** Stator and rotor currents of the motor with speed-controlled starting.

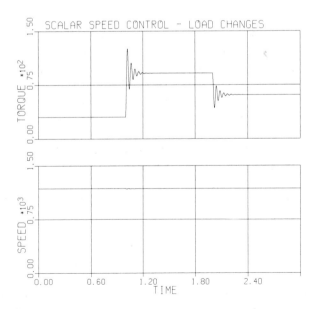

**Figure 2.24.** Torque and speed of the motor with rapid load changes.

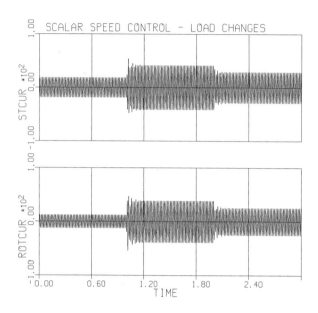

**Figure 2.25.** Stator and rotor currents of the motor with rapid load changes.

### Simulation 2.3. Scalar Speed Control - Reversing

To illustrate further the dynamic performance of a speed-controlled induction motor during radical changes in operating conditions, reversing of the example drive system is simulated. The same system as in Simulation 2.1, running under the steady state conditions with speed of 1183 rpm, is subjected to a step change of the reference speed to -1183 rpm. The control system forces the motor to operate in the generating mode which causes the drive system to decelerate. When the reference frequency, $f^*$, reaches zero, switching signals for the power switches in phases B and C of the inverter are interchanged, causing the magnetic field of the stator to reverse its direction of rotation. In this way, negative $f^*$ is translated into a negative phase sequence from the inverter and a resultant negative speed of the stator field.

The torque and speed of the motor are shown in Figure 2.26. It can be seen that at the new speed command the motor slows down rapidly, then accelerates in the opposite direction. The torque waveform then becomes similar to that at starting (see Figure 2.22). The negative torque and positive speed during deceleration confirm that the motor operates as a generator. Stator and rotor current waveforms are shown in Figure 2.27.

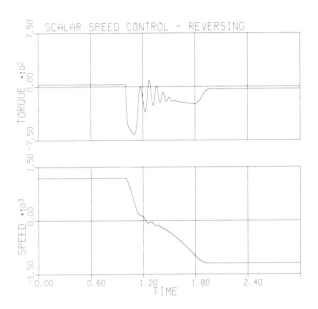

**Figure 2.26.** Torque and speed of the motor during reversing.

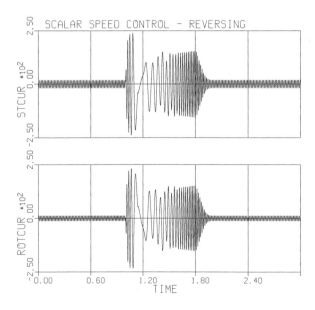

**Figure 2.27.** Stator and rotor currents of the motor during reversing.

## Simulation 2.4. Torque-Controlled Starting

To illustrate the operation of the steady-state torque control system described in Section 2.6, torque-controlled starting of the example motor is simulated. Continuous control of the stator current is assumed, i.e., both the magnitude and frequency are adjusted every 0.1 msec by the microprocessor-based control system. The load is the same as in the previous simulations.

The goal is to reach the speed of 1183 rpm within 1 sec from the starting instant and to maintain this speed later on. Consequently, the reference torque, $T^*$, is set to 129.7 N m for $t \leq 1$ sec and 30.6 N m for t > 1 sec. The rotor flux, $\Lambda_R'$, is to be maintained at its rated level of 0.3 Wb/ph, as calculated in Example 2.5.

The motor torque and speed are shown in Figure 2.28. The motor operates in a transient state when accelerated from start, then settles to a steady state when the developed torque matches the load torque. As seen from the torque waveform, the system is not able to maintain the instantaneous torque at the desired, fixed, level during the transient state. However, the average torque follows the reference value approximately so that the speed reached in the predicted time of 1 sec is close to the target value of 1183 rpm.

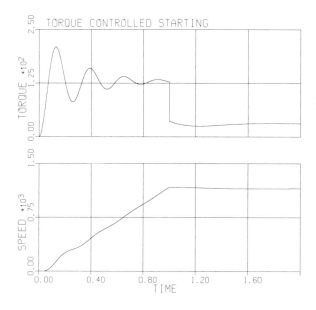

**Figure 2.28.** Torque and speed of the motor with torque-controlled starting.

# SCALAR CONTROL OF INDUCTION MOTORS

The same applies to the rotor flux and the angle between vectors $i_R^e$ and $\lambda_R^e$, which is displayed for reference in further considerations, is shown in Figure 2.29. The flux stabilizes after a relatively long period of time, due to the large time constant of the rotor. Since the settling time for first-order dynamic systems equals four time constants, then, with $\tau_r = 0.267$ sec, the settling time is about 1.1 sec. The angle between the current and flux vectors, oscillates around the value of 90° during the transient state, settling down to this value in the steady state. The settling time is the same as that of the rotor flux.

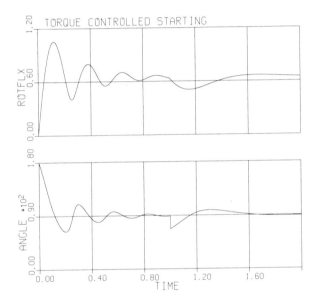

**Figure 2.29.** Rotor flux of the motor and angle between vectors $i_R^e$ and $\lambda_R^e$ during torque-controlled starting.

It must be stressed that the rotor flux illustrated is the magnitude, $\lambda_R$, of the vector $\lambda_R^e$ and this differs from the rms value, $\Lambda_R'$, of the rotor flux per phase in the Γ' equivalent circuit, which was used as one of the inputs to the control system. Specifically, according to Eqs. (1.60) and (2.38),

$$\lambda_R = \frac{1.5\sqrt{2}}{\gamma'}\Lambda_R' \qquad (2.73)$$

which, in the case considered, yields $\lambda_R = 0.65$ Wb. It should also be mentioned that the model of an induction motor used in this book does not take account of saturation of the magnetic circuit of the motor. Therefore, in a real machine, the amplitude of oscillations of the flux and torque would be smaller than those shown here.

The stator and rotor currents are shown in Figure 2.30. The gradual increase of frequency during acceleration of the motor is easily observed.

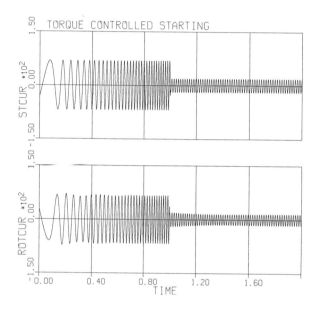

**Figure 2.30.** Stator and rotor currents of the motor with torque-controlled starting.

### Simulation 2.5. Scalar Torque Control - Reversing

This simulation illustrates the dynamic performance of a torque-controlled motor during reversing. The example motor runs at a speed of 1000 rpm with the same load as in the previous simulations, and is forced to reach a speed of -1000 rpm within 1.5 sec. This requires a motor torque of -81.1 N m for 0.75 sec to decelerate the system to zero, a torque of -142.3 N m for the next 0.75 sec to accelerate in the opposite direction to the target speed, and a torque of -30.6 N m from then on to maintain this speed. As in the case of the speed-controlled motor in Simulation 2.3, the negative supply frequency is realized by changing the phase sequence of the inverter output.

The resultant torque and speed of the motor are shown in Figure 2.31. During the reversing, the instantaneous torque oscillates around the around the corresponding reference values and stabilizes in the steady state after completion of the operation. In effect, the speed curve, particularly in the first stage of the reversing, does not follow the assumed straight line exactly and does not attain the precise target value, settling down at -959 rpm.

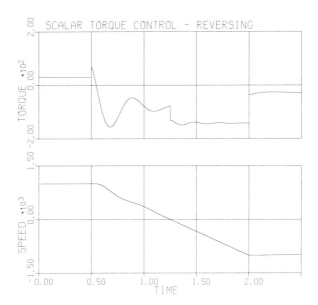

**Figure 2.31.** Torque and speed of the motor with torque-controlled reversing.

The rotor flux and the angle between vectors $i_R^e$ and $\lambda_R^e$ are shown in Figure 2.32. Large swings of these variables can be observed before the flux stabilizes again at the reference level and the angle changes from the initial value of 90° to the final value of -90°. As in Simulation 2.4, the flux oscillations are unrealistically high since saturation of the iron of the motor has not been taken into account.

Another idealization used in this and the previous simulations, namely the assumption of instantaneous current control, results in apparent discontinuities in certain quantities, discernible on some of the presented plots. In reality, current-controlled inverters have a finite, although short, response time, so that no instantaneous changes in the motor variables take place.

Stator and rotor current waveforms are shown in Figure 2.33. Note the saddle-type transient during the transition from a positive to a negative speed.

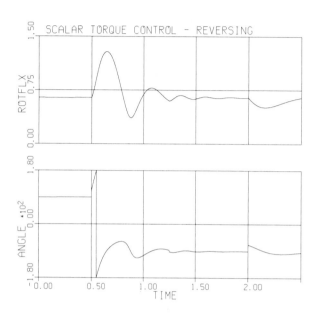

**Figure 2.32.** Rotor flux of the motor and angle between vectors $i_R^e$ and $\lambda_R^e$ during torque-controlled reversing.

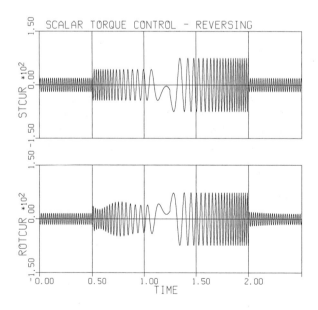

**Figure 2.33.** Stator and rotor currents of the motor with torque-controlled reversing.

## Chapter 3

## FIELD ORIENTATION PRINCIPLE

In general, an electric motor can be thought of as a controlled source of torque. Accurate control of the instantaneous torque produced by a motor is required in high-performance drive systems, e.g., those used for position control. The torque developed in the motor is a result of the interaction between current in the armature winding and the magnetic field produced in the field system of the motor. The field should be maintained at a certain optimal level, sufficiently high to yield a high torque per unit ampere, but not too high to result in excessive saturation of the magnetic circuit of the motor. With fixed field, the torque is proportional to the armature current.

Independent control of the field and armature currents is feasible in separately-excited d.c. motors where the current in the stator field winding determines the magnetic field of the motor, while the current in the rotor armature winding can be used as a direct means of torque control. The physical disposition of the brushes with respect to the stator field ensures optimal conditions for torque production under all conditions. Even today, most high-performance drive systems are still based on d.c. motors.

In a similar manner to that in d.c. machines, in induction motors the armature winding is also on the rotor, while the field is generated by currents in the stator winding. However, the rotor current is not directly derived from an external source but results from the e.m.f. induced in the winding as a result of the relative motion of the rotor conductors with respect to the stator field. In other words, the stator current is the source of both the magnetic field and armature current. In the most commonly used, squirrel-cage motors, only the stator current can be directly controlled, since the rotor winding is not accessible. Optimal torque production conditions are not inherent due to the absence of a fixed physical disposition between the stator and rotor fields, and the torque equation is nonlinear. In effect, independent and efficient control of the field and torque is not as simple and straightforward as in d.c. motors.

The concept of steady-state torque control of an induction motor, described in Chapter 2, is extended to transient states of operation in the high-performance, vector-controlled a.c. drive systems based on the Field Orientation Principle (FOP). The FOP defines conditions for decoupling the field control from the torque control. A field-oriented induction motor emulates a separately-excited d.c. motor in two aspects:

(1) Both the magnetic field and the torque developed in the motor can be controlled independently.

(2) Optimal conditions for torque production, resulting in the maximum torque per unit ampere, occur in the motor both in the steady state and in transient conditions of operation.

## 3.1 Optimal Torque Production Conditions

Optimal conditions for electromagnetic torque production in a current-carrying coil exposed to an externally-generated magnetic field occur when the coil plane is parallel to the lines of the field, i.e., when the vector of the coil current, $i$, is orthogonal to the vector, $\lambda$, of flux produced by the external source and linking the coil. Although the coil produces its own mmf, yet, as pointed out in Section 1.4, the resultant component of the total flux does not affect the torque.

Non-optimal and optimal torque production conditions are illustrated in Figure 3.1. The orthogonality of vectors $i$ and $\lambda$ gives the electrodynamic forces, $F$, the best leverage, thereby resulting in maximum torque. The same conclusion could have been drawn from Figure 1.8 and Eq. (1.41).

As mentioned before, the optimal torque production conditions are inherently satisfied in a d.c. motor, illustrated in Figure 3.2. Brushes supplying the armature current, $i_a$, to the rotor winding, via a commutator, are positioned in such a way that the armature current vector, $i_a$, is always orthogonal to the flux vector (field flux), $\lambda_f$, produced in the stator and linking the rotor winding. In effect, the developed torque, $T$, is proportional both to the armature current and the field flux, i.e.,

$$T = k_T i_a \lambda_f \quad (3.1)$$

# FIELD ORIENTATION PRINCIPLE

where $k_T$ is a constant depending on the physical parameters of the motor.

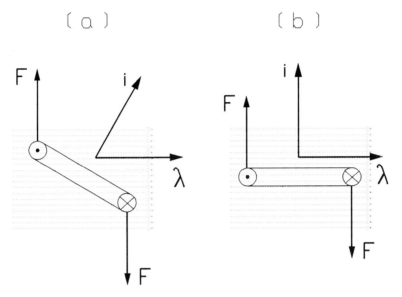

**Figure 3.1.** Non-optimal (a) and optimal (b) torque production conditions.

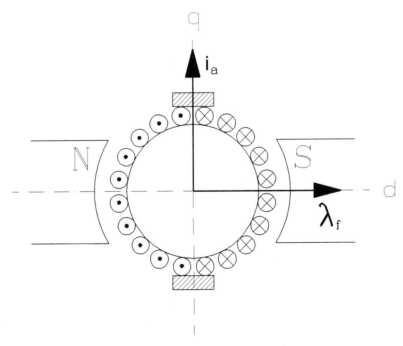

**Figure 3.2.** Torque production in a d.c. motor.

Since in a separately-excited d.c. motor the armature current, $i_a$, and field flux, $\lambda_f$, can be controlled independently, then, according to Eq. (3.1), the motor can be thought of as a linear current-to-torque converter with an adjustable gain $k_T \lambda_f$. The block diagram representation of such a motor is shown in Figure 3.3. Although the diagram is analogous to that of the torque-controlled induction motor, shown in Figure 2.9, the signals involved represent instantaneous values of the motor variables, while those in the induction-motor diagram are r.m.s., steady-state values. The goal of the FOP is to extend such a representation of an induction motor to instantaneous values as well.

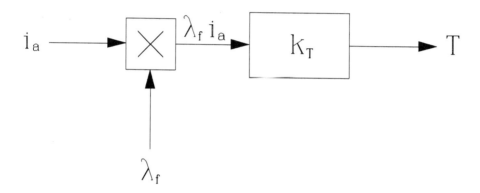

**Figure 3.3.** Block diagram representation of a separately-excited d.c. motor.

## 3.2 Dynamic Block Diagram of an Induction Motor in the Excitation Reference Frame

The torque equation (1.90) in the excitation frame can be rearranged so that the torque is expressed in terms of the stator current and rotor flux. From Eq. (1.87),

$$\mathbf{1}_R^e = \frac{1}{L_r}(\boldsymbol{\lambda}_R^e - L_m \mathbf{1}_S^e) . \qquad (3.2)$$

Substituting Eq. (3.2) in Eq. (1.90) and making use of the rotor time constant, $\tau_r = L_r/R_r$, introduced already in Eq. (2.52), the torque equation can be written as

# FIELD ORIENTATION PRINCIPLE

$$T = \frac{P}{3R_r} \frac{L_m}{\tau_r} (i_{QS}^e \lambda_{DR}^e - i_{DS}^e \lambda_{QR}^e) \ . \tag{3.3}$$

To construct a block diagram of the motor, in which components $i_{DS}^e$ and $i_{QS}^e$ of the stator current vector, $\mathbf{i}_S^e$, represent the input variables and the torque, $T$, the output variable, the rotor flux vector components, $\lambda_{DR}^e$ and $\lambda_{QR}^e$, appearing in Eq. (3.3) must be expressed in terms of $i_{DS}^e$ and $i_{QS}^e$.

As $v_R^e = 0$ (shorted rotor winding), Eq. (1.85) can be written as

$$R_r \mathbf{i}_R^e + (p+j\omega_r) \boldsymbol{\lambda}_R^e = 0 \tag{3.4}$$

and, after substituting Eq. (3.2) for $\mathbf{i}_R^e$ and utilizing the rotor time constant, $\tau_r$, rearranged to

$$p\boldsymbol{\lambda}_R^e = \frac{1}{\tau_r} [L_m \mathbf{i}_S^e - (1+j\omega_r \tau_r) \boldsymbol{\lambda}_R^e] \ . \tag{3.5}$$

Resolving the vectors appearing in Eq. (3.5) into their $DQ$ components and dividing the resultant equations by $p$ yields

$$\lambda_{DR}^e = \frac{1}{p} (\frac{L_m}{\tau_r} i_{DS}^e - \frac{1}{\tau_r} \lambda_{DR}^e + \omega_r \lambda_{QR}^e) \tag{3.6}$$

$$\lambda_{QR}^e = \frac{1}{p} (\frac{L_m}{\tau_r} i_{QS}^e - \frac{1}{\tau_r} \lambda_{QR}^e - \omega_r \lambda_{DR}^e) \ . \tag{3.7}$$

Eqs. (3.3), (3.6), and (3.7) describe the dynamic block diagram of an induction motor, shown in Figure 3.4. The diagram illustrates the source of difficulties in the control of induction motors. Besides the four multipliers making the motor a nonlinear system, there are two cross-couplings between the $D$ and $Q$ paths. As shown later, the FOP represents a special case of a known method of nonlinear decoupling used in the control of nonlinear systems. The nonlinear decoupling consists in controlling selected variables in such a way that they are always equal to zero. In effect, certain parts of the mathematical model become irrelevant and can then be removed from the model.

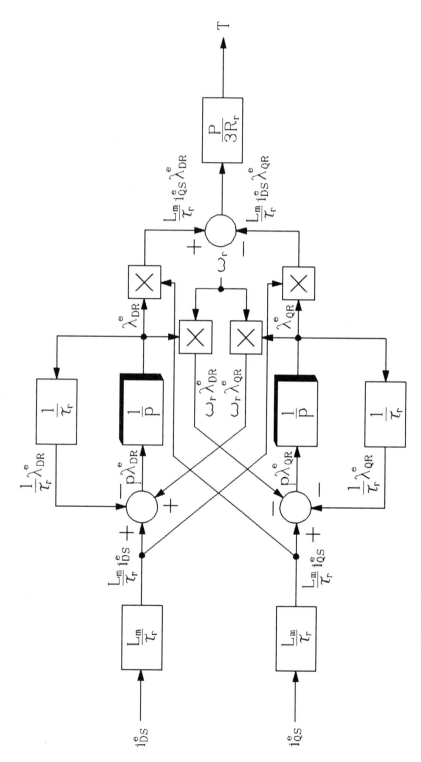

Figure 3.4. Dynamic block diagram for an induction motor in the excitation frame.

# FIELD ORIENTATION PRINCIPLE

## 3.3 Field Orientation Conditions

It follows from Eq. (3.3) that if

$$\lambda_{QR}^e = 0 \qquad (3.8)$$

then

$$T = \frac{P}{3R_r}\frac{L_m}{\tau_r}\lambda_{DR}^e i_{QS}^e = k_T \lambda_{DR}^e i_{QS}^e \qquad (3.9)$$

where

$$k_T = \frac{P}{3R_r}\frac{L_m}{\tau_r} = \frac{P}{3}\frac{L_m}{L_r}. \qquad (3.10)$$

Clearly, Eq. (3.9) is analogous to Eq. (3.1) describing a separately-excited d.c. motor. Hence, when Eq. (3.8) is satisfied and

$$\lambda_{DR}^e = const \qquad (3.11)$$

then an induction motor represents a linear current-to-torque converter.

Inspection of the block diagram of an induction motor in Figure 3.4 reveals that setting $\lambda_{QR}^e$ to zero in accordance with Eq. (3.8) allows significant reduction of the diagram. Such a reduced diagram is shown in Figure 3.5. Further reduction is possible by replacing the $i_{DS}^e \rightarrow \lambda_{DR}^e$ path by a single, first-order dynamic block having a transfer function

$$G(p) = \frac{\lambda_{DR}^e}{i_{DS}^e} = \frac{L_m}{\tau_r}\frac{\frac{1}{p}}{1+\frac{1}{p}\times\frac{1}{\tau_r}} = \frac{L_m}{\tau_r p + 1}. \qquad (3.12)$$

The resulting block diagram of a field-oriented induction motor is depicted in Figure 3.6. Independent control of flux and torque has now been achieved. While the torque response to a change of current $i_{QS}^e$ is instantaneous, the response of flux $\lambda_{DR}^e$ to changes of current $i_{DS}^e$ is inertial, with time constant $\tau_r$.

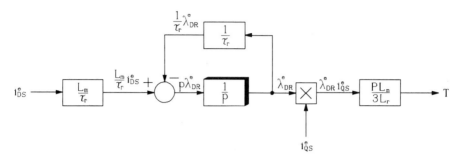

**Figure 3.5.** Block diagram of an induction motor with $\lambda_{QR}^e = 0$.

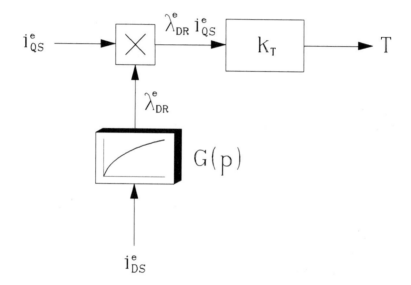

**Figure 3.6.** Block diagram of a field-oriented induction motor.

From Eq. (3.4),

$$i_{DR}^e = \frac{1}{R_r}(\omega_r \lambda_{QR}^e - p\lambda_{DR}^e). \qquad (3.13)$$

If conditions (3.8) and (3.11) are satisfied, i.e., $\lambda_{QR}^e = 0$ and $p\lambda_{DR}^e = 0$, then $i_{DR}^e = 0$, i.e., $i_R^e = ji_{QR}^e$. At the same time, $\lambda_R^e = \lambda_{DR}^e$. Consequently, vectors $i_R^e$ and $\lambda_R^e$ are orthogonal which, as shown in Figure 3.7, represents optimal conditions for the torque production, analogous to those illustrated in Figure 3.2 for a d.c. motor. In recognition of the influence of the angle

FIELD ORIENTATION PRINCIPLE                                          95

between vectors $i_R^e$ and $\lambda_R^e$ on the developed torque, this angle will subsequently be called a torque angle.

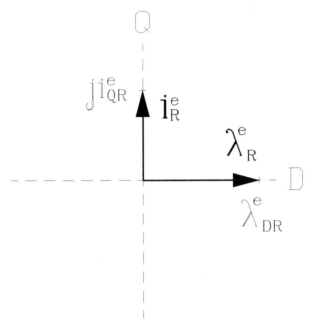

**Figure 3.7.** Vectors of rotor current and flux in a field-oriented induction motor.

In an induction motor, the optimal torque-production conditions are always satisfied in the steady state. The $\Gamma'$ steady-state equivalent circuit indicates orthogonality of phasors $\hat{I}_R'$ and $\hat{I}_M'$, the latter being proportional to the rotor flux phasor, $\hat{\Lambda}_R'$. Hence, vectors $i_R^{e'}$ and $\lambda_R^{e'}$ and, consequently, vectors $i_R^e$ and $\lambda_R^e$, are also orthogonal. This has been confirmed in Simulations 2.4 and 2.5, as seen in Figures 2.29 and 2.32. Independent control of torque and flux is realized in the scalar torque control system described in Sections 2.5 and 2.6. In effect, a motor operating in that system can be considered field-oriented, but only in the steady state.

# Chapter 4

## CLASSIC FIELD ORIENTATION SCHEMES

As demonstrated in Chapter 3, the FOP defines conditions for optimal torque production and the decoupling of torque control from field control, both under steady-state and transient operating conditions of an induction motor. Orthogonality of the rotor current and flux vectors must be maintained at all times. This requirement is inherently satisfied in the steady state when the rotor settles down to such a speed that the developed torque matches the load torque. Under transient conditions, however, in order to meet the FOP conditions, special techniques are required to provide an algorithmic equivalent of the actual physical disposition between the stator and rotor fields of the emulated d.c. motor.

The difference between a field-oriented induction motor and a d.c. machine can be illustrated by a simple analogy to a broom balanced in the vertical position on a finger tip, versus the same broom firmly held in a fist. In both cases, the broom, which can be thought of as the vector of armature current, is required to maintain a 90° angle with respect to the arm which represents the vector of armature flux. Assuming an ideal steady-state situation, i.e., a perfectly steady hand, perfectly linear, constant-speed motion of the holder, a perfectly balanced broom, and an absence of external forces such as aerodynamic drag, the broom would rest vertically on the finger tip and move with it, even though not otherwise stabilized. However, under transient conditions, e.g., when changing speed, balancing the broom in the constantly orthogonal orientation to the arm, in this analogy to an induction motor, requires considerable skill and continuous, quick reaction of the holder. In contrast, in an analogy to a d.c. motor, if the broom is firmly held in the fist, it can easily be made to maintain that orientation under all conditions.

In this chapter, basic field orientation schemes for induction motors are described. Two, often called "classic", approaches to vector control are introduced: a direct method, in which sensors are used to measure the magnetic field of the motor, and an indirect method, based on measurement of the rotor position. In both of these, the rotating, excitation reference frame is aligned with the vector of rotor flux.

## 4.1 Field Orientation with Respect to the Rotor Flux Vector

The general block diagram of a vector control system for an induction motor is shown in Figure 4.1. A field orientation system produces reference signals, $i_{as}^*$, $i_{bs}^*$, and $i_{cs}^*$, of the stator currents, based on the input reference values, $\lambda_r^*$ and $T^*$, of the rotor flux and motor torque, respectively, and signals corresponding to selected variables of the motor. An inverter supplies the motor currents, $i_{as}$, $i_{bs}$, and $i_{cs}$, such that their waveforms follow the reference waveforms, $i_{as}^*$, $i_{bs}^*$, and $i_{cs}^*$.

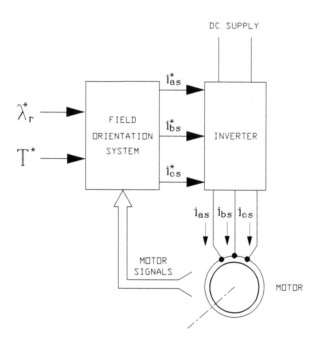

**Figure 4.1.** General block diagram of a vector control system for an induction motor.

As shown in Figure 3.6, in a field-oriented induction motor, the $i_{DS}^e$ and $i_{QS}^e$ components of the stator current vector, $i_S^e$, in the excitation frame can be used for independent control of the motor field and torque, respectively.

## CLASSIC FIELD ORIENTATION SCHEMES

Hence, the field orientation system shown in Figure 4.1 first converts $\lambda_r^*$ and $T^*$ into the corresponding reference signals, $i_{DS}^{e*}$ and $i_{QS}^{e*}$, of the vector of stator current, and then transforms these into the reference signals, $i_{as}^*$, $i_{bs}^*$, and $i_{cs}^*$, of the stator phase currents which are to be produced by the inverter.

To understand the operating algorithm of the field orientation system, it should be noticed that the reference stator currents, $i_{as}^*$, $i_{bs}^*$, and $i_{cs}^*$, can easily be calculated using the $dq \rightarrow abc$ transformation, if the corresponding reference signals, $i_{ds}^{s*}$ and $i_{qs}^{s*}$, in the stator reference frame are known. This is a simple scalar, or static, transformation, since the elements of the transformation matrix used to perform this operation are constants (see Eq. (1.14)).

However, it can be seen from Eq. (1.81) that a dynamic transformation, i.e., one involving time, is required to determine $i_{ds}^{s*}$ and $i_{qs}^{s*}$ from $i_{DS}^{e*}$ and $i_{QS}^{e*}$. Yet, Eq. (1.81) cannot be used directly since it is impossible to say what specific values should be substituted for $\omega$ and $t$. The problem is that the excitation frame has been defined as rotating with the same angular velocity as the vector quantities of the motor, but no assumptions have been made with respect to synchronization of the frame with these vectors. Clearly, any one of the vectors can be used as a reference with which the excitation frame is to be aligned. In fact, various field orientation techniques employ various reference vectors. In the classic approach described, it is the rotor flux vector, $\lambda_r^s$, along which the excitation frame is oriented. Hence, this method is usually referred to as the rotor flux orientation scheme.

Denoting the angular position of the rotor flux vector in the stator reference frame by $\Theta_r$, the $DQ \rightarrow dq$ transformation in the described scheme is expressed as

$$\begin{bmatrix} i_{ds}^{s*} \\ i_{qs}^{s*} \end{bmatrix} = \begin{bmatrix} \cos(\Theta_r) & -\sin(\Theta_r) \\ \sin(\Theta_r) & \cos(\Theta_r) \end{bmatrix} \begin{bmatrix} i_{DS}^{e*} \\ i_{QS}^{e*} \end{bmatrix}. \quad (4.1)$$

The two reference frames synchronized by this means, together with the rotor flux vector used as a reference vector, are shown in Figure 4.2. It can be observed that this orientation of the excitation frame inherently satisfies the FOP condition (3.18). The motor then conforms to the block diagram representation in Figure 3.6. The rotor flux is controlled by adjusting the $i_{DS}^e$ component of the stator current vector, independently from the torque control, which is realized by means of the $i_{QS}^e$ component.

The only requirement for this scheme is an accurate identification of angle $\Theta_r$, i.e., the current position of vector $\lambda_r^s$. This can be done in either a direct or indirect way, as described in the next two sections.

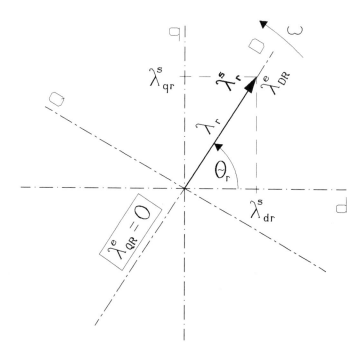

**Figure 4.2.** Orientation of the excitation reference frame along the rotor flux vector.

## 4.2 Direct Rotor Flux Orientation Scheme

In direct field orientation systems, the magnitude and angular position (phase) of the reference flux vector are either measured or estimated from the stator voltage and currents using flux observers. Hall sensors can, for instance, be employed to measure magnetic fields. Placing the sensors in the air gap of the motor, in the $d$ and $q$ axes, allows determination of the corresponding components of vector $\lambda_m^s$ of the mutual flux (airgap flux). In the T dynamic equivalent circuit of an induction motor, shown in Figure 4.3, a derivative $p\lambda_m^s$ of this flux appears across the mutual inductance, $L_m$. Hence,

$$\lambda_m^s = L_m \mathbf{i}_m^s = L_m(\mathbf{i}_s^s + \mathbf{i}_r^s) \qquad (4.2)$$

# CLASSIC FIELD ORIENTATION SCHEMES

or

$$\mathbf{i}_r^s = \frac{1}{L_m}\boldsymbol{\lambda}_m^s - \mathbf{i}_s^s. \qquad (4.3)$$

Since $\lambda_r^s$ differs from $\lambda_m^s$ by only the leakage flux in the rotor, then

$$\begin{aligned}
\boldsymbol{\lambda}_r^s &= \boldsymbol{\lambda}_m^s + L_{lr}\mathbf{i}_r^s = \boldsymbol{\lambda}_m^s + L_{lr}\left(\frac{1}{L_m}\boldsymbol{\lambda}_m^s - \mathbf{i}_s^s\right) \\
&= \left(1 + \frac{L_{lr}}{L_m}\right)\boldsymbol{\lambda}_m^s - L_{lr}\mathbf{i}_s^s = \frac{L_r}{L_m}\boldsymbol{\lambda}_m^s - L_{lr}\mathbf{i}_s^s.
\end{aligned} \qquad (4.4)$$

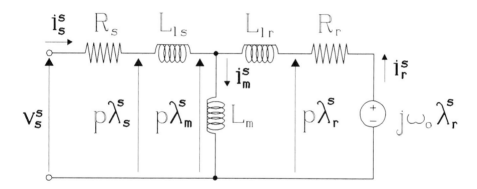

**Figure 4.3.** Dynamic T equivalent circuit of an induction motor.

A microprocessor-based rotor flux calculator is shown in Figure 4.4. It performs the following algebraic operations:

(1) Signals $i_{ds}^s$ and $i_{qs}^s$ are calculated from the actual stator currents, $i_{as}$, $i_{bs}$, and $i_{cs}$, using the $abc \rightarrow dq$ transformation expressed by Eq. (1.11).

(2) Using Eq. (4.4), signals $\lambda_{dr}^s$ and $\lambda_{qr}^s$ are calculated as

$$\lambda_{dr}^s = \frac{L_r}{L_m}\lambda_{dm}^s - L_{lr}i_{ds}^s \qquad (4.5)$$

$$\lambda_{qr}^s = \frac{L_r}{L_m}\lambda_{qm}^s - L_{1r}i_{qs}^s. \quad (4.6)$$

(3) Magnitude, $\lambda_r$, and phase, $\Theta_r$, of the rotor flux vector are determined using the rectangular to polar coordinate transformation

$$\lambda_{dr}^s + j\lambda_{qr}^s \rightarrow \lambda_r \angle \Theta_r. \quad (4.7)$$

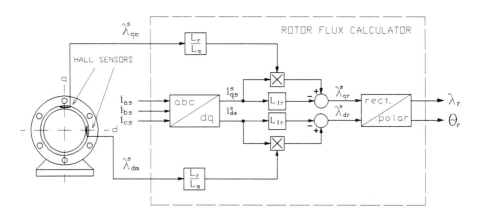

**Figure 4.4.** Determination of the magnitude and position of the rotor flux vector using Hall sensors and a rotor flux calculator.

In practice, the ratio of $L_r$ to $L_m$ and the rotor leakage inductance, $L_{1r}$, are not significantly affected by changes in the operating conditions of the motor, such as the winding temperature or saturation of the magnetic circuit. Therefore, the field orientation technique described is considered to be the most robust and accurate. However, it requires the placement of vulnerable Hall sensors in the air gap of the motor, to the detriment of the cost and reliability of the drive system.

It must be pointed out that the orthogonal spacing of the flux sensors in Figure 4.4 applies only to 2-pole motors. In a $P$-pole motor, the sensors must be placed $180°/P$ from each other.

Since $\lambda_{DR}^e = \lambda_r$ (see Figure 4.2), then the output variable, $\lambda_r$, of the rotor flux calculator can be used as a feedback signal in the field control loop. The same variable can also be utilized for calculation of the developed

# CLASSIC FIELD ORIENTATION SCHEMES

torque to obtain a feedback signal for the torque control loop. A torque calculator, illustrated in Figure 4.5, computes the developed torque in the following steps:

(1) The static $abc \rightarrow dq$ transformation is performed on the stator currents, $i_{as}$, $i_{bs}$, and $i_{cs}$, to obtain $i^s_{ds}$ and $i^s_{qs}$.

(2) Angle $\Theta_r$, supplied by the rotor flux calculator is substituted in Eq. (1.80) for $\omega t$ in order to transform signals $i^s_{ds}$ and $i^s_{qs}$ into the $i^e_{QS}$ component of the vector of stator current in the excitation frame.

(3) Magnitude, $\lambda_r$, of the rotor flux, also supplied by the rotor flux calculator and presumed equal to $\lambda^e_{DR}$, is multiplied by $i^e_{QS}$ and by the torque constant, $k_T$, given by Eq. (3.20), to calculate the developed torque from Eq. (3.19).

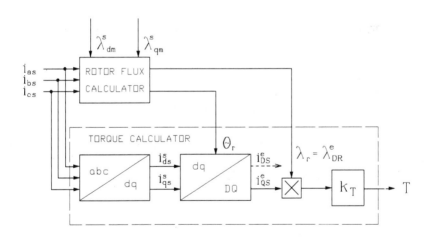

**Figure 4.5.** Torque calculator.

The block diagram of a vector control system for an induction motor employing the direct rotor flux orientation scheme described above is shown in Figure 4.6. The system corresponds to the general block diagram of a field-oriented motor in Figure 3.6, where $i^e_{DS}$ and $i^e_{QS}$ constitute the flux-producing and torque-producing currents, respectively. In the system in Figure 4.7, the reference currents, $i^{e*}_{DS}$ and $i^{e*}_{QS}$, in the excitation frame are aligned with the rotor flux vector, $\lambda^s_r$, by using the angle $\Theta_r$ signal provided by the rotor flux calculator to the $DQ \rightarrow dq$ transformation block. Reference currents, $i_{ds}^{s*}$ and $i_{qs}^{s*}$, in the stator frame, are calculated in this block using Eq. (4.1), and converted into reference stator currents, $i_{as}^{*}$, $i_{bs}^{*}$, and $i_{cs}^{*}$, in the $dq \rightarrow abc$ transformation block. Following the reference waveforms, the inverter supplies currents $i_{as}$, $i_{bs}$, and $i_{cs}$ to the stator of the controlled motor.

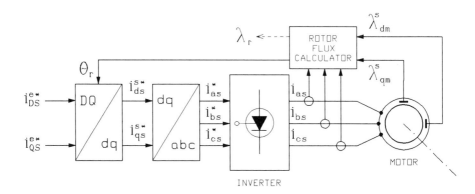

**Figure 4.6.** Basic direct rotor flux orientation system.

The unused output signal, $\lambda_r$, of the rotor flux calculator can be employed for closed-loop field control, while the torque calculator in Figure 4.5 may provide a torque feedback signal. The basic system in Figure 4.6, when augmented in this way, allows independent, closed-loop control of both the field and torque of the motor. The resultant vector control system of an induction motor, with proportional-plus-integral (PI) field and torque controllers, is depicted in Figure 4.7.

In a similar manner to that used in the scalar torque control system described in Section 2.6, a current-controlled, voltage-source inverter is assumed as a source of stator currents in the systems in Figures 4.6 and 4.7. In practice, current-source inverters are also used, especially in high-power drive systems. However, since these are controlled somewhat differently from voltage-source inverters, and in order to concentrate the attention of the reader on the field orientation issues, only the current-controlled, voltage-source type of inverter is considered here. This inverter is capable of producing low-distortion currents of any arbitrary waveform, hence it can be assumed that the actual currents, $i_{as}$, $i_{bs}$, and $i_{cs}$, in the stator closely reflect the reference current signals, $i_{as}^{*}$, $i_{bs}^{*}$, and $i_{cs}^{*}$, generated in the field orientation system.

# CLASSIC FIELD ORIENTATION SCHEMES

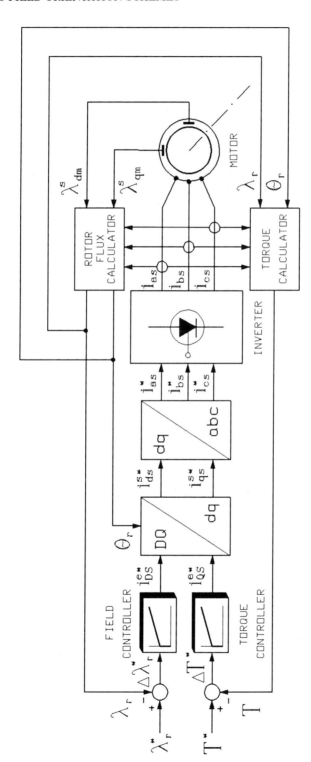

**Figure 4.7.** Vector control system for an induction motor with direct rotor flux orientation.

## 4.3 Indirect Rotor Flux Orientation Scheme

In the direct rotor flux orientation scheme presented in the preceding section, the angular position, $\Theta_r$, of the rotor flux vector, necessary for the $DQ \to dq$ transformation, is measured directly, using airgap flux sensors. The indirect approach is, instead, based on the calculation of the slip speed, $\omega_r^*$, required for correct field orientation, and the imposition of this speed on the motor.

If the synchronous speed necessary to maintain the orthogonal orientation of vectors $\lambda_R^e$ and $i_R^e$ in the given operating conditions of the motor is denoted by $\omega^*$, the $\Theta_r$ angle can be expressed as

$$\Theta_r = \int_0^t \omega^* dt = \int_0^t \omega_r^* dt + \int_0^t \omega_o dt \qquad (4.8)$$
$$= \int_0^t \omega_r^* dt + \Theta_o$$

where $\Theta_o$ is the angular displacement of the rotor, which is easy to measure using a shaft position sensor.

The required value of the slip speed, $\omega_r^*$, can be computed from the equations of the motor under field orientation conditions. Since $\lambda_R^e = \lambda_{DR}^e$, then Eq. (3.2) becomes

$$i_R^e = \frac{1}{L_r} (\lambda_{DR}^e - L_m i_s^e) . \qquad (4.9)$$

Substituting Eq. (4.9) in the rotor equation (3.4) gives

$$\lambda_{DR}^e [1 + \tau_r (p + j\omega_r)] = L_m i_s^e . \qquad (4.10)$$

The real and imaginary parts of Eq. (4.10) are

$$\lambda_{DR}^e (1 + \tau_r p) = L_m i_{DS}^e \qquad (4.11)$$

and

# CLASSIC FIELD ORIENTATION SCHEMES

$$\omega_r \tau_r \lambda^e_{DR} = L_m i^e_{QS}. \qquad (4.12)$$

Replacing $\omega_r$ with $\omega_r^*$, $\lambda^e_{DR}$ with $\lambda_r^*$, and $i^e_{QS}$ with $i^{e*}_{QS}$ in the last equation, and solving for $\omega_r^*$, yields

$$\omega_r^* = \frac{L_m}{\tau_r} \frac{i^{e*}_{QS}}{\lambda_r^*}. \qquad (4.13)$$

It can be shown that, unsurprisingly, Eq. (4.13) constitutes a time-domain equivalent of the phasor-domain equation (2.53). Indeed, from Eq. (4.11), in the steady state of the motor ($p = 0$),

$$\lambda_r^* = L_m i^{e*}_{DS} \qquad (4.14)$$

which, when substituted in Eq. (4.13), gives

$$\omega_r^* = \frac{1}{\tau_r} \frac{i^{e*}_{QS}}{i^{e*}_{DS}}. \qquad (4.15)$$

Variables $i^{e*}_{DS}$ and $i^{e*}_{QS}$ represent the required flux-producing and torque-producing components of the stator current vector, $\vec{i}_s^*$, respectively, in the same way as $I^*_{s\Phi}$ and $I^*_{sT}$ in Eq. (2.53) are the respective components of the stator current phasor, $\vec{I}_s$.

Signal $i^{e*}_{DS}$, corresponding to a given reference rotor flux, $\lambda_r^*$, can be determined from Eq. (4.11), as

$$i^{e*}_{DS} = \frac{1 + \tau_r p}{L_m} \lambda_r^* \qquad (4.16)$$

while signal $i^{e*}_{QS}$ for a given reference torque, $T^*$, can be obtained from the torque equation (3.9) of a field-oriented motor, as

$$i^{e*}_{QS} = \frac{T^*}{k_T \lambda_r^*} \qquad (4.17)$$

where the torque constant, $k_T$, is given by Eq. (3.10). Since the control scheme presented constitutes an extension of the scalar torque control method, described in Section 2.6, on transient states of operation of the motor, the reference flux and torque values must satisfy the SOAR conditions (2.58) and (2.63).

A vector control system for an induction motor based on the indirect rotor flux orientation scheme is shown in Figure 4.8. The rotor flux and developed torque are controlled in a feedforward manner, i.e., no feedback loops are employed. As a consequence of this, performance of the system strongly depends on an accurate knowledge of motor parameters, a requirement that is difficult to satisfy in practical applications. On the other hand, a major advantage of such a system is that a standard motor can be used, whose rotor position is easily measurable by an external sensor.

The phase angle, $\Theta_r$, of the rotor flux vector is calculated by integrating the reference slip speed, $\omega_r^*$, and adding the resultant angle, $\Theta^*$, to the rotor angular displacement, $\Theta_o$, in accordance with Eq. (4.8). As implied by Eq. (4.13), the rotor time constant, $\tau_r$, plays a crucial role in the proper field orientation. Unfortunately, the time constant varies significantly during operation of an induction motor, mainly because of changes in the rotor resistance with temperature, but also due to frequency-dependent skin effects in the rotor conductors and the impact of magnetic saturation on the rotor inductance. In general, the indirect field orientation scheme can be evaluated as being simpler than the direct field orientation one, but this simplicity is paid for by a penalty in the form of poorer performance.

## 4.4 Examples and Simulations

The examples and simulations presented in this section are intended to illustrate the concept of field orientation with respect to the rotor flux vector and to demonstrate the operation of the direct and indirect vector control systems. For simplicity, the inverter supplying the motor is assumed to be an ideal source of stator currents, i.e., $i_{as} = i_{as}^*$, $i_{bs} = i_{bs}^*$, and $i_{cs} = i_{cs}^*$.

### Example 4.1.

The example motor operates in the steady state under rated conditions. Utilize results of Example 1.2 to calculate the airgap flux and rotor flux vectors from the dynamic T equivalent circuit of the motor. Use the results so obtained to validate Eqs. (4.5) and (4.6) of the rotor flux calculator.

# CLASSIC FIELD ORIENTATION SCHEMES

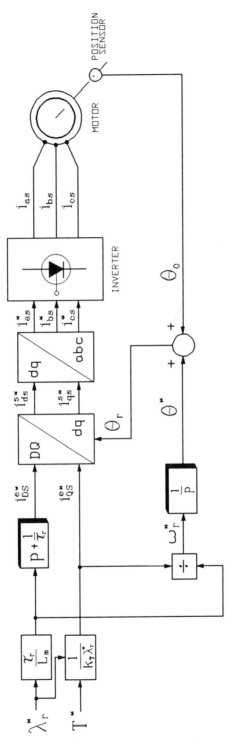

**Figure 4.8.** Vector control system for an induction motor with indirect rotor flux orientation.

## Solution

From Example 1.2,

$$\mathbf{i}_s^s = 50.5 e^{j(377t-25.5°)} \text{ A}$$

$$\mathbf{i}_r^s = 47.1 e^{j(377t+172.8°)} \text{ A}.$$

Hence,

$$\mathbf{i}_m^s = \mathbf{i}_s^s + \mathbf{i}_r^s = 50.5 e^{j(377t-25.5°)} + 47.1 e^{j(377t+172.8°)}$$
$$= 15.9 e^{j(377t-94.2°)} \text{ A}$$

and

$$\boldsymbol{\lambda}_m^s = L_m \mathbf{i}_m^s = 0.041 \times 15.9 e^{j(377t-94.2°)}$$
$$= 0.6511 e^{j(377t-94.2°)} \text{ Wb}.$$

The rotor flux vector can now be calculated as

$$\boldsymbol{\lambda}_r^s = \boldsymbol{\lambda}_m^s + L_{lr} \mathbf{i}_r^s = 0.6511 e^{j(377t-94.2°)}$$
$$+ 0.00074 \times 47.1 e^{j(377t+172.8°)}$$
$$= 0.6496 e^{j(377t-97.2°)} \text{ Wb}.$$

The airgap flux and rotor flux differ very little, which is understandable in view of the low value of $L_{lr}$ implying little leakage of flux in the magnetic circuit of the rotor. Note that alignment of the excitation frame with the vector of stator voltage has been assumed in Example 1.2 and maintained in the current example. It can also be seen that the vectors of rotor flux and rotor current are orthogonal, as

$$\angle(\boldsymbol{\lambda}_r^s, \mathbf{i}_r^s) = -97.2° - 172.8° = -270° = 90°.$$

This agrees with the observation made in Section 3.3 that optimal torque-production conditions occur during steady-state operation of an induction motor.

The *dq* components of the vectors of stator current and airgap flux are

$$i_{ds}^s = 50.5\cos(377t-25.5°) \text{ A}$$

$$i_{qs}^s = 50.5\sin(337t-25.5°) \text{ A}$$

$$\lambda_{dm}^s = 0.6511\cos(377t-94.2°) \text{ Wb}$$

$$\lambda_{qm}^s = 0.6511\sin(377t-94.2°) \text{ Wb}.$$

Hence, according to Eqs. (4.5) and (4.6),

$$\begin{aligned}\lambda_{dr}^s &= \frac{0.0417}{0.041} \times 0.6511\cos(377t-94.2°) \\ &\quad -0.00074 \times 50.5\cos(377t-25.5°) \\ &= 0.6496\cos(377t-97.2°) \text{ Wb}\end{aligned}$$

and

$$\begin{aligned}\lambda_{qr}^s &= \frac{0.0417}{0.041} \times 0.6511\sin(377t-94.2°) \\ &\quad -0.00074 \times 50.5\sin(377t-25.5°) \\ &= 0.6496\sin(377t-97.2°) \text{ Wb}.\end{aligned}$$

Consequently,

$$\boldsymbol{\lambda_r^s} = \lambda_{dr}^s + j\lambda_{qr}^s = 0.6496[\cos(377t-97.2°) + j\sin(377t-97.2°)] = \mathbf{0.6496}e^{j(377t-97.2°)} \text{ Wb}.$$

The last result, identical with that obtained previously, confirms the feasibility of determination of the magnitude and phase of the rotor flux vector from the measurements of stator currents and the spatial components of the airgap flux.

## Example 4.2.

The example motor operates in the direct field orientation system described in Section 4.3 under the same conditions as in the previous example. Show the computations performed by the flux and torque calculators following the sampling of the airgap flux and stator currents at $t = 5$ msec.

### Solution

The values of the stator currents sampled in the system at the given instant of time can be determined from the phasor of the stator current found in Example 1.1 as $\hat{I}_s = 23.8 \angle -25.5°$ A/ph. The value 23.8 A/ph is an r.m.s. magnitude of the phasor, and not an amplitude of the stator phase currents. Hence, at $\omega t = 377$ rad/sec $\times$ 0.005 sec = 1.885 rad = 108°,

$$i_{as} = \sqrt{2} \times 23.8 \cos(108° - 25.5°) = 4.4 \text{ A}$$

$$i_{bs} = \sqrt{2} \times 23.8 \cos(108° - 25.5° - 120°) = 26.7 \text{ A}$$

$$i_{cs} = \sqrt{2} \times 23.8 \cos(108° - 25.5° - 240°) = -31.1 \text{ A}$$

while the $dq$ components of the airgap flux vector can be determined using the results of the preceding example, as,

$$\lambda_{dm}^s = 0.6511 \cos(108° - 94.2°) = 0.6323 \text{ Wb}$$

$$\lambda_{qm}^s = 0.6511 \sin(108° - 94.2°) = 0.1553 \text{ Wb}.$$

Employing the $abc \rightarrow dq$ transformation, the $dq$ components of the stator current vector are computed by the flux calculator as

$$\begin{bmatrix} i_{ds}^s \\ i_{qs}^s \end{bmatrix} = \begin{bmatrix} 1 & -\frac{1}{2} & -\frac{1}{2} \\ 0 & \frac{\sqrt{3}}{2} & -\frac{\sqrt{3}}{2} \end{bmatrix} \begin{bmatrix} 4.4 \\ 26.7 \\ -31.1 \end{bmatrix} = \begin{bmatrix} 6.6 \\ 50.1 \end{bmatrix} \text{ A}.$$

# CLASSIC FIELD ORIENTATION SCHEMES

In the next step, Eqs. (4.5) and (4.6) are used to yield

$$\lambda_{dr}^s = \frac{0.0417}{0.041} \times 0.6323 - 0.00074 \times 6.6 = 0.6382 \text{ Wb}$$

$$\lambda_{qr}^s = \frac{0.0417}{0.041} \times 0.1553 - 0.00074 \times 50.1 = 0.1209 \text{ Wb}.$$

Finally, the magnitude and phase of the rotor flux vector are calculated as

$$\lambda_r = \sqrt{\lambda_{dr}^{s\,2} + \lambda_{qr}^{s\,2}} = \sqrt{0.6382^2 + 0.1209^2} = \mathbf{0.6496 \text{ Wb}}$$

and, since $\lambda_{dr}^s \neq 0$,

$$\Theta_r = \tan^{-1}\left(\frac{\lambda_{qr}^s}{\lambda_{dr}^s}\right) + [sgn(\lambda_{dr}^s) - 1] \times 90°$$
$$= \tan^{-1}\left(\frac{0.1209}{0.6382}\right) + (1-1) \times 90° = \mathbf{10.8°}.$$

The results so obtained can be confirmed by substituting 108° for 377*t* in the expression for $\lambda_r^s$ determined in the preceding example.

The torque calculator first performs the same *abc→dq* transformation of stator currents as does the flux calculator (clearly, in a practical system, this transformation can be done once for both the calculators). Then, the torque developed in the motor is calculated from Eq. (3.9) with $\lambda_r$ substituted for $\lambda_{DR}^e$. In the case under consideration, the excitation reference frame is aligned with the vector of stator voltage and not with the rotor flux vector. Nevertheless, the vectors of the rotor current and flux are orthogonal, as shown in the preceding example, hence the motor can be considered as field-oriented, only with respect to the stator voltage vector and not the rotor flux vector. Therefore the value of $i_{QS}^e$ in the existing excitation reference frame can safely be used. This value is calculated from Eq. (4.1) as

$$i_{QS}^e = -6.6\sin(10.8°) + 50.1\cos(10.8°) = \mathbf{48.0 \text{ A}}$$

and, since

$$k_T = \frac{6}{3} \times \frac{0.041}{0.0417} = 1.966 \ N\ m/Wb/A$$

then

$$T = 1.966 \times 0.6496 \times 48.0 = \mathbf{61.2\ N\ m}.$$

The calculated torque agrees with that determined in Examples 1.1 and 1.2 involving the same motor under the same operating conditions.

### Example 4.3.

Illustrate the insensitivity of the direct rotor flux orientation scheme to the variation of motor parameters by assuming that the flux and torque calculators utilize incorrect values of the inductance coefficients. Take $L_m$ as 20% higher than the actual mutual inductance and $L_{lr}$ as 40% higher than the actual rotor leakage inductance, and calculate the flux and torque feedback signals corresponding to those in the preceding example. Compare the results.

### Solution

Since $L_m = 1.2 \times 0.041 = 0.0492$ H and $L_{lr} = 1.4 \times 0.00074 = 0.00104$ H, then $L_r = 0.0492 + 0.00104 = 0.05024$ H. Hence,

$$\lambda_{dr}^s = \frac{0.05024}{0.0492} \times 0.6323 - 0.00104 \times 6.6 = 0.6388\ Wb$$

and

$$\lambda_{qr}^s = \frac{0.05024}{0.0492} \times 0.1553 - 0.00104 \times 50.1 = 0.1065\ Wb$$

i.e.,

$$\lambda_r = \mathbf{0.6476\ Wb}$$

$$\Theta_r = \mathbf{9.5°}.$$

CLASSIC FIELD ORIENTATION SCHEMES

Comparing these values with those of 0.6496 Wb and 10.8° obtained in Example 4.2, it can be seen that the magnitude error of the flux calculator is only 0.3% while the angle error is 12%.

The torque calculator computes $i_{QS}^e$ as

$$i_{QS}^e = -6.6 \sin(9.5°) + 50.1 \cos(9.5°) = 48.3 \text{ A}.$$

Since the torque constant in this case is

$$k_T = \frac{6}{3} \times \frac{0.0492}{0.05024} = 1.959 \text{ N m/Wb/A}$$

then

$$T = 1.959 \times 0.6476 \times 48.3 = \mathbf{61.3 \ N \ m}.$$

In spite of the relatively high angle error of the flux calculator, the error of the torque calculator is less than 0.2%. In a practical system, such an accuracy would be considered ideal in comparison with that of the other elements of the system, such as the flux sensors or inverter.

### Example 4.4.

The example motor operates in the vector control system based on the indirect rotor flux orientation, shown in Figure 4.8. The flux and torque commands represent the rated values of the rotor flux and motor torque, i.e., $\lambda_r^* = 0.6496$ Wb and $T^* = 61.2$ N m. Follow the operation of the system to calculate the required stator currents, $i_{as}^*$, $i_{bs}^*$, and $i_{cs}^*$, at the instant of time $t = 0.5$ sec since starting the motor. Assume that the rotor has turned 10 rad since starting.

*Solution*

Signals $i_{DS}^{e*}$ and $i_{QS}^{e*}$ are calculated from Eqs. (4.16) and (4.17), as

$$i_{DS}^{e*} = \frac{0.6496}{0.041} = 15.84 \text{ A}$$

$$i_{qs}^{e*} = \frac{61.2}{1.966 \times 0.6496} = 47.92 \text{ A}.$$

The reference slip speed, $\omega_r^*$, is determined from Eq. (4.13) as

$$\omega_r^* = \frac{0.041}{0.267} \times \frac{47.92}{0.6496} = 11.33 \text{ rad/sec}.$$

The integral, $\Theta^*$, of $\omega_r^*$ at $t = 0.5$ sec is

$$\Theta^* = \omega_r^* t = 11.33 \times 0.5 = 5.665 \text{ rad}$$

while the rotor angle, $\Theta_o$, of an equivalent 2-pole motor equals $P/2$ of the angular displacement of the actual motor, i.e.,

$$\Theta_o = \frac{6}{2} \times 10 = 30 \text{ rad}.$$

Consequently, angle $\Theta_r$ for the $DQ \rightarrow dq$ transformation is

$$\Theta_r = \Theta^* + \Theta_o = 5.665 + 30 = 35.665 \text{ rad}$$

which is equivalent to 243.5°.

From Eq. (4.1),

$$\begin{bmatrix} i_{ds}^{s*} \\ i_{qs}^{s*} \end{bmatrix} = \begin{bmatrix} \cos(243.5°) & -\sin(243.5°) \\ \sin(243.5°) & \cos(243.5°) \end{bmatrix} \begin{bmatrix} 15.84 \\ 47.92 \end{bmatrix}$$

$$= \begin{bmatrix} 35.82 \\ -35.56 \end{bmatrix} \text{ A}$$

hence, from Eq. (1.15),

$$\begin{bmatrix} i_{as}^* \\ i_{bs}^* \\ i_{cs}^* \end{bmatrix} = \begin{bmatrix} \frac{2}{3} & 0 \\ -\frac{1}{3} & \frac{1}{\sqrt{3}} \\ -\frac{1}{3} & -\frac{1}{\sqrt{3}} \end{bmatrix} \begin{bmatrix} 35.82 \\ -35.56 \end{bmatrix} = \begin{bmatrix} 23.88 \\ -32.47 \\ 8.59 \end{bmatrix} \text{ A.}$$

Analogous computations are repeated within every sampling interval of the digital control system of the motor.

### Simulation 4.1. Direct Rotor Flux Orientation System

To illustrate the performance of the direct rotor flux orientation system, the example motor is simulated under programmed-torque operating conditions. The load torque is 30.6 N m and the mass moment of inertia of the load is 0.1 kg m². The motor is to accelerate from standstill to a speed of 1000 rpm in 0.5 sec, run with this speed for the next 0.5 sec, and then reverse, to reach a speed of -1000 rpm within 1 sec. This requires the following torque program to be realized:

$$T^* = \begin{cases} 135.3 \ N \ m \ \text{for} & 0 < t \leq 0.5 & \text{sec} \\ 30.6 \ N \ m \ \text{for} & 0.5 < t \leq 1.0 & \text{sec} \\ -74.1 \ N \ m \ \text{for} & 1.0 < t \leq 1.5 & \text{sec} \\ -135.3 \ N \ m \ \text{for} & 1.5 < t \leq 2.0 & \text{sec} \\ -30.6 \ N \ m \ \text{for} & t > 2.0 & \text{sec.} \end{cases}$$

The rotor flux is to be maintained at 0.5 Wb.

The inverter is assumed to be an ideal controlled current source. Since the motor is current-driven, the stator voltages are not known and the current equation (1.35) cannot be used. Therefore, for computation of the rotor flux and current, Eqs. (1.52) and (1.54) have been rearranged to

$$\frac{d\pmb{\lambda}_r^s}{dt} = \frac{1}{\tau_r} [(j\omega_o \tau_r - 1) \pmb{\lambda}_r^s + L_m \pmb{i}_s^s] \qquad (4.18)$$

and

$$\mathbf{i}_r^s = \frac{1}{L_r} (\boldsymbol{\lambda}_r^s - L_m \mathbf{i}_s^s) . \qquad (4.19)$$

After resolution into the *dq* components, the following equations were obtained and used in the numerical simulation program:

$$\frac{d\lambda_{dr}^s}{dt} = \frac{1}{\tau_r} (-\lambda_{dr}^s - \omega_o \tau_r \lambda_{qr}^s + L_m i_{ds}^s) \qquad (4.20)$$

$$\frac{d\lambda_{qr}^s}{dt} = \frac{1}{\tau_r} (\omega_o \tau_r \lambda_{dr}^s - \lambda_{qr}^s + L_m i_{qs}^s) \qquad (4.21)$$

$$i_{dr}^s = \frac{1}{L_r} (\lambda_{dr}^s - L_m i_{ds}^s) \qquad (4.22)$$

$$i_{qr}^s = \frac{1}{L_r} (\lambda_{qr}^s - L_m i_{qs}^s) . \qquad (4.23)$$

The torque and speed of the motor are shown in Figure 4.9, the stator and rotor currents in Figure 4.10, and the rotor flux and torque angle in Figure 4.11. It can be seen that the desired torque, speed, and rotor flux programs are perfectly realized. The orthogonality of rotor flux and current vectors is maintained all the time (note that the torque angle changes its sign with the change in the torque direction). Obviously, such ideal performance of the system is partly due to the idealized model of the inverter, the assumed perfect accuracy of sensors, and knowledge of exact motor parameters. However, the robustness of this scheme, illustrated in Example 4.3, allows real systems to operate with a similar high level of performance.

### Simulation 4.2. Indirect Rotor Flux Orientation System (I)

The same motor, but in the indirect rotor flux orientation system was simulated. An accurate knowledge of parameters of the motor was assumed. The results, shown in Figures 4.12 through 4.14 are identical with those obtained in the preceding simulation.

# CLASSIC FIELD ORIENTATION SCHEMES 119

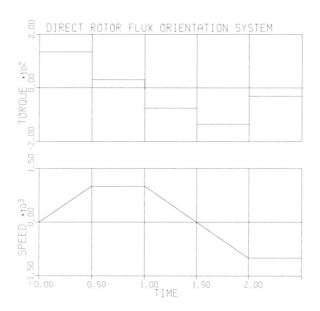

**Figure 4.9.** Torque and speed of the motor in the direct rotor flux orientation system.

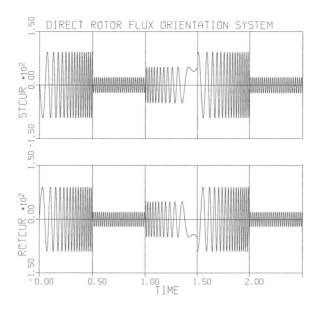

**Figure 4.10.** Stator and rotor currents of the motor in the direct rotor flux orientation system.

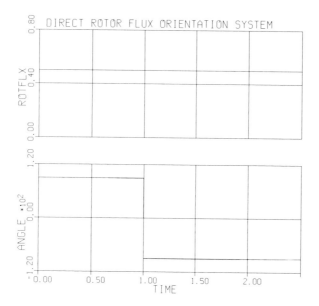

**Figure 4.11.** Rotor flux and torque angle of the motor in the direct rotor flux orientation system.

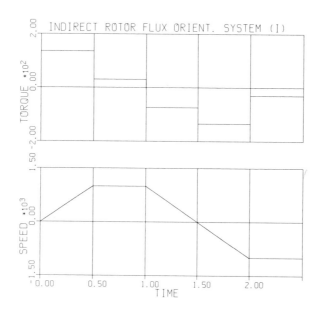

**Figure 4.12.** Torque and speed of the motor in the indirect rotor flux orientation system with accurate estimation of motor parameters.

# CLASSIC FIELD ORIENTATION SCHEMES

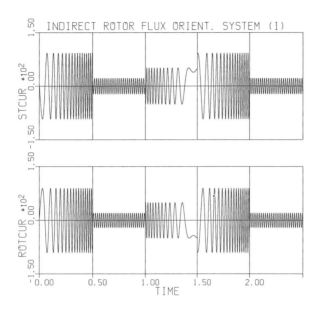

**Figure 4.13.** Stator and rotor currents of the motor in the indirect rotor flux orientation system with accurate estimation of motor parameters.

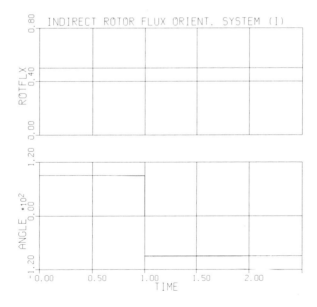

**Figure 4.14.** Rotor flux and torque angle of the motor in the indirect rotor flux orientation system with accurate estimation of motor parameters.

### Simulation 4.3. Indirect Rotor Flux Orientation System (II)

Vector control systems with indirect field orientation are inherently less robust that those employing the direct (feedback) approach. To illustrate this problem, it is assumed that the inductance parameters of the motor used in the settings of the control system are overestimated by 10%, while the rotor resistance is underestimated, also by 10%. This is typical of the situation when the effects of saturation of the magnetic circuit of the motor and of the temperature of the windings are not properly taken into account.

The operating conditions of the motor are the same as in the previous simulations. The torque and speed of the motor are shown in Figure 4.15, stator and rotor currents in Figure 4.16, and the rotor flux and torque angle in Figure 4.17.

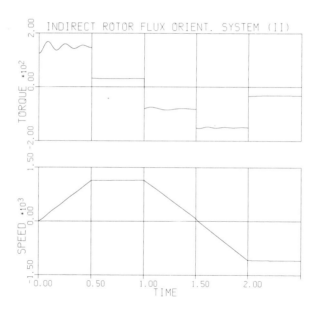

**Figure 4.15.** Torque and speed of the motor in the indirect rotor flux orientation system with inaccurate estimation of motor parameters.

It can be observed that the motor performance is significantly poorer in comparison with that in Simulation 4.2. The developed torque, rotor flux, and torque angle are characterized by oscillatory transients. The average torque and rotor flux are higher than the required values by over 10%. In effect, the steady-state speeds are also higher than the target values.

# CLASSIC FIELD ORIENTATION SCHEMES

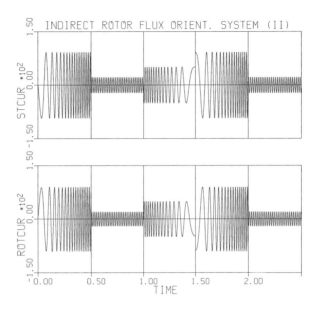

**Figure 4.16.** Stator and rotor currents of the motor in the indirect rotor flux orientation system with inaccurate estimation of motor parameters.

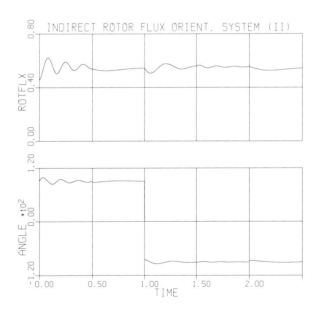

**Figure 4.17.** Rotor flux and torque angle of the motor in the indirect rotor flux orientation system with inaccurate estimation of motor parameters.

## Chapter 5

## INVERTERS

Three-phase inverters, supplying voltages and currents of adjustable frequency and magnitude to the stator, are an important element of adjustable speed drive systems employing induction motors. Therefore, a comprehensive treatment of their operating principles and characteristics is presented in this chapter.

Inverters, based on semiconductor power switches, are d.c. to a.c. static power converters. Depending on the type of d.c. source supplying the inverter, they can be classified as voltage source inverters (VSI) or current source inverters (CSI). In practice, the d.c. source is usually a rectifier, typically of the three-phase bridge configuration, with the so-called d.c. link connected between the rectifier and the inverter. The d.c. link is a simple inductive, capacitive, or inductive-capacitive low-pass filter. Since neither the voltage across a capacitor nor the current through an inductor can change instantaneously, a capacitive-output d.c. link is used for a VSI and an inductive-output link is employed in a CSI. In battery-powered drive systems, such as those for electric vehicles, the rectifier is, obviously, not needed. However, the d.c. link is still used as an interface, either to impose the current-source input to a CSI, or to protect the battery from the high-frequency component of the supply current of a VSI.

VSIs can be either voltage or current-controlled. In a voltage-controlled inverter, it is the frequency and magnitude of the fundamental of the output voltage that is adjusted. Feedforward voltage control is employed, since the inverter voltage is dependent only on the supply voltage and the states of the inverter switches, and, therefore, accurately predictable. Current-controlled VSIs require sensors of the output currents which provide the necessary control feedback. Voltage-controlled VSIs are mostly used in scalar speed-control systems based on the CVH principle described in Sections 2.2 and 2.3, while current-controlled VSIs and CSIs are typical for torque controlled drives, with either scalar (Sections 2.5 and 2.6) or vector (Chapter 4) control.

The type of semiconductor power switch used in an inverter depends on the volt-ampere rating of the inverter, as well as on other operating and economic considerations, such as the switching frequency or cost of the system. Modern power electronics has emerged as an important field of electrical engineering since the invention of thyristors in late fifties. Subsequently, many other semiconductor power devices have appeared on the market, gradually replacing classic thyristors (SCRs), primarily in low and medium power converters. A variety of devices, such as the gate turn-off thyristor (GTO), bipolar junction power transistor (BJT), power MOSFET, insulated gate bipolar transistor (IGBT), static induction transistor (SIT) and thyristor (SITH), and MOS-controlled thyristor (MCT), are now available to design engineers. What is probably the most popular power switch nowadays, the IGBT, appears in most of low and medium power inverters in commercial practice, while GTOs, with volt-ampere ratings equalling those of SCRs, are becoming increasingly used in high power applications.

With the exception of SCRs, now disappearing from inverter circuits, all the other devices mentioned above are fully controlled switches which can be turned both on and off by appropriate switching signals. In the latter parts of this chapter, generic, ideal switches are assumed. Such idealization consists of assumptions of unlimited switching frequency and zero switching loss, zero voltage drop across the switch in the ON-state, and zero current through the switch in the OFF-state.

## 5.1 Voltage Source Inverter

A diagram of the power circuit of a three-phase VSI is shown in Figure 5.1. The circuit has bridge topology with three branches (phases), each consisting of two power switches and two freewheeling diodes. In the case illustrated, the inverter is supplied from an uncontrolled, diode-based rectifier, via a d.c. link which contains an LC filter in the inverted Γ configuration. Whilst this circuit represents a standard arrangement, it allows only positive power flow, i.e., from the supply system, typically a three-phase power line, to the load. Negative power flow, which occurs when the load feeds power back to the supply, is not possible since the resulting negative d.c. component of the current in the d.c. link cannot pass through the rectifier diodes. Therefore, in drive systems where the VSI-fed motor may operate as a generator, more complex supply systems must be used. These involve either a braking resistance connected across the d.c.

# INVERTERS

link, or replacement of the uncontrolled rectifier by a so-called dual converter. If the power returned by the motor is dissipated in a braking resistor, the process is called dynamic braking. The dual converter allows the power from the d.c. link to flow back into the supply system, resulting in the more efficient, regenerative braking of the motor. Sometimes, a number of inverter-motor pairs are supplied from a single rectifier and d.c. link. In this case, if the power flow from the supply system is always positive, i.e., the power produced by some of the motors is fully consumed by the remaining motors, the basic supply arrangement, shown in Figure 5.1., can still be employed. In battery supplied VSIs, there is, clearly, no problem with negative power flow.

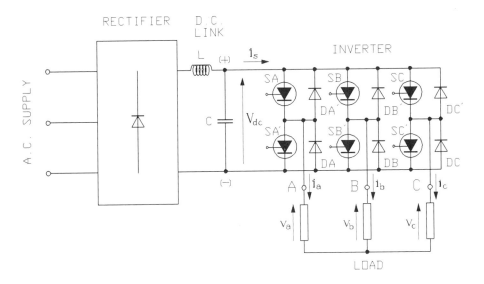

**Figure 5.1.** Circuit diagram of a three-phase VSI.

In the circuit shown in Figure 5.1, the power switches in a given branch must never both be in the ON-state, since this would constitute a short circuit. On the other hand, if both the switches are in the OFF-state, then the potential of the corresponding output terminal is unknown to the control system of the inverter because no voltage sensors are employed. The circuit can be completed through either the upper or lower diode and, consequently, the potential can be equal to that of either the positive bus, (+), or negative bus, (-). Therefore, the inverter is controlled in such a way that, in a given branch, either the upper switch (e.g., SA) is ON and the lower switch (e.g., SA') is OFF or, vice-versa, the upper switch is OFF and the lower switch is ON.

Considering, for instance, branch A of the inverter, if the output current, $i_a$, is positive, as in Figure 5.1, and switch SA is ON, the current flows from the positive bus, through the switch, to the load. If the switch is turned off and, simultaneously, switch SA' is turned on, the current is forced to flow from the negative bus and through diode DA, since the inductive nature of the load impedance prevents the current from a rapid change in direction which would then allow it to flow through the freshly turned on switch SA'. Conversely, a negative current in phase A can either flow through switch SA', when in the ON-state, or through diode DA' when the switch is OFF. It can be seen, that a given phase current flowing through either switch is associated with the flow of electric power from the supply source to the load, while the current flowing through either freewheeling diode is associated with the flow of power in the opposite direction, specifically, to the capacitor C in the d.c. link.

Since only two combinations of states of the switches in each branch are allowed, a switching (logic) variable can be assigned to each phase of the inverter. In effect, only eight logic states are permitted for the whole power circuit. Defining the switching variables as

$$a = \begin{cases} 0 & \text{if SA is OFF and SA' is ON} \\ 1 & \text{if SA is ON and SA' is OFF} \end{cases} \quad (5.1)$$

$$b = \begin{cases} 0 & \text{if SB is OFF and SB' is ON} \\ 1 & \text{if SB is ON and SB' is OFF} \end{cases} \quad (5.2)$$

$$c = \begin{cases} 0 & \text{if SC is OFF and SC' is ON} \\ 1 & \text{if SC is ON and SC' is OFF} \end{cases} \quad (5.3)$$

the instantaneous values of the line-to-line output voltages of the inverter are given by

$$V_{ab} = V_{dc}(a-b) \quad (5.4)$$

$$V_{bc} = V_{dc}(b-c) \quad (5.5)$$

$$V_{ca} = V_{dc}(c-a) \quad (5.6)$$

INVERTERS

where $V_{dc}$ is the d.c. supply voltage of the inverter.

In balanced three-phase systems, the line-to-neutral voltages can be calculated from the line-to-line voltages as

$$V_a = \frac{1}{3}(V_{ab} - V_{ca}) \qquad (5.7)$$

$$V_b = \frac{1}{3}(V_{bc} - V_{ab}) \qquad (5.8)$$

$$V_c = \frac{1}{3}(V_{ca} - V_{bc}). \qquad (5.9)$$

Hence, after substituting Eqs. (5.4) through (5.6) in Eqs. (5.7) through (5.9), the line-to-neutral voltages of the inverter are given by

$$V_a = \frac{V_{dc}}{3}(2a - b - c) \qquad (5.10)$$

$$V_b = \frac{V_{dc}}{3}(2b - c - a) \qquad (5.11)$$

$$V_c = \frac{V_{dc}}{3}(2c - a - b). \qquad (5.12)$$

It follows from Eqs. (5.4) through (5.6) and (5.10) through (5.12) that line-to-line voltages can assume only three values: $-V_{dc}$, 0, and $V_{dc}$, while line-to-neutral voltages can assume five values: $-2/3\ V_{dc}$, $-1/3\ V_{dc}$, 0, $1/3\ V_{dc}$, and $2/3\ V_{dc}$. The eight logic states of the inverter can be numbered from 0 to 7 using the decimal equivalent of the binary number $abc_2$. For example, if $a = 1$, $b = 0$, and $c = 1$, then $abc_2 = 101_2 = 5_{10}$ and the inverter is said to be in state 5. Taking $V_{dc}$ as the base voltage, the per-unit output voltages are: $v_{ab} = 1$, $v_{bc} = -1$, $v_{ca} = 0$, $v_a = 1/3$, $v_b = -2/3$, and $v_c = 1/3$.

Performing the $abc \rightarrow dq$ transformation, the output voltages can be represented as space vectors in the stator reference frame, each vector corresponding to a given state of the inverter. The space vector diagrams of line-to-line and line-to-neutral voltages of a VSI are shown in Figure 5.2. The vectors are presented in the per-unit format. The line-to-line voltage vectors are denoted by upper-case letters and the line-to-neutral voltage vectors by lower-case letters.

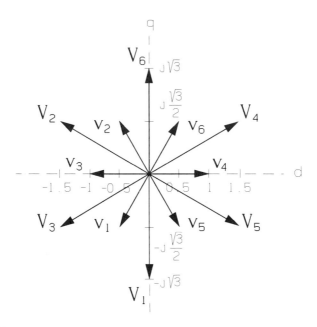

**Figure 5.2.** Space vectors of output voltages of a VSI.

## 5.2 Voltage Control in Voltage Source Inverters

A voltage-controlled VSI can be thought of as a sequential ring counter which follows a certain sequence of logic states with a period equal to that of the desired output frequency. If the state sequence is 4 - 6 - 2 - 3 - 1 - 5 ..., with the duration of each state equal to one sixth of the period of the output frequency, then the output voltage waveforms are as shown in Figure 5.3. It can be seen, that although these waveforms are stepped, they constitute coarse approximations of the corresponding waveforms of an ideal, balanced, three-phase, sinusoidal a.c. system. This, the so-called square-wave operation mode is, by far, the simplest, and is characterized by very low switching losses in the inverter switches. Within a cycle of the output frequency, each switch is turned on and off only once, hence even very slow switches can be used. Consequently, this mode of operation is frequently encountered in SCR-based VSIs, now becoming obsolete and phased out of industrial practice.

It is not possible to control the magnitude of the output voltages in inverters operating in the square-wave mode. In these, an adjustable

# INVERTERS

source of d.c. supply voltage is therefore required, and a controlled, SCR-based, rectifier, or a chopper, is used for this purpose.

The harmonic spectrum of the line-to-line voltage of a square-wave VSI is shown in Figure 5.4, with the relative r.m.s. magnitudes of the harmonics expressed in dB. In this case, the r.m.s. value of the fundamental equals $\sqrt{6}/\pi \approx 0.78$ of the d.c. supply voltage, and represents the highest achievable d.c. to a.c. voltage conversion ratio of all the various types of inverters. Even and triplen harmonics are absent due to the half-wave symmetry of the waveforms, balanced operation, and three-wire configuration of the inverter. The lowest-order higher harmonic is the fifth. Harmonics of the output currents are attenuated approximately in proportion to their order by the load inductances, hence current waveforms are closer to ideal sinusoids than are the voltage waveforms. Even so, a significant harmonic content ("ripple") is present in the currents, and square-wave operation is used only when there is no other viable technical alternative, e.g., in low-performance, high-power drives.

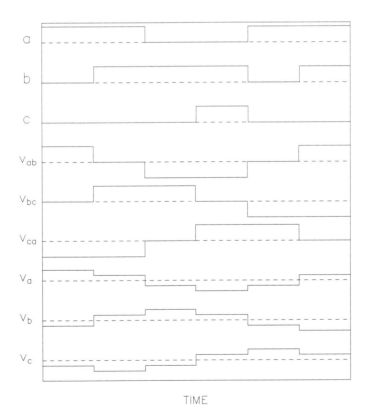

**Figure 5.3.** Switching signals and output voltages of a VSI in the square-wave operating mode.

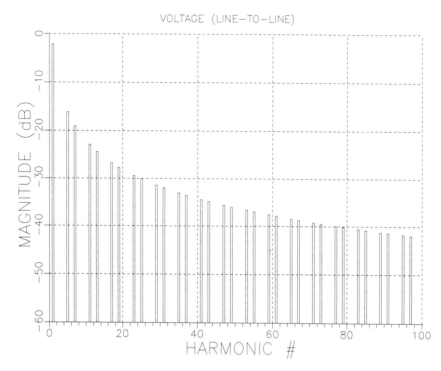

**Figure 5.4.** Harmonic spectrum of the line-to-line voltage of a VSI in the square-wave operating mode.

Simultaneous control of the frequency and magnitude of the output voltages is performed by means of the so-called pulse width modulation (PWM). It can be observed in Figure 5.3 that in the square-wave operating mode the switching variables consist of single pulses within a cycle of the output frequency. If, instead of this, multiple pulses with varying, appropriately modulated widths are used to control the inverter switches, a significant improvement in the harmonic spectra of the output voltage and current waveforms may be achieved. Pulse width modulation also allows to control the magnitude of the fundamental voltage from zero to a specific maximum value which depends on the d.c. supply voltage of the inverter.

A large number of different PWM techniques have been developed and implemented in practical inverters. Currently, one of the most popular methods is based on the concept of space vectors of the inverter voltages, already shown in Figure 5.2. In later considerations it will be assumed that the inverter load is a wye connected induction motor. Consequently, the load currents are generated by the line-to-neutral voltages of the inverter, and it is control of these voltages which will be focused on.

# INVERTERS

Space vectors of the line-to-neutral voltages are shown again in Figure 5.5, together with an arbitrary reference vector, $v^*$, to be generated by the inverter. Note that in addition to the six non-zero vectors produced in states 1 through 6 of the inverter, two zero vectors, $v_0$ and $v_7$ are also indicated on the diagram. These correspond to states 0 and 7, when all the output voltages are zero due to clamping of the output terminals of the inverter to either the positive or negative supply bus. Clearly, only vectors $v_0$ through $v_7$, further referred to as base vectors, can be produced at a given instant of time. Therefore, vector $v^*$ is represents an average rather than an instantaneous value, the average being taken over a period of the so-called switching, or sampling, interval which, in practice, constitutes a small fraction of the cycle of output frequency. The switching interval, in the center of which the reference vector is located, is shown in Figure 5.5 as the shaded segment.

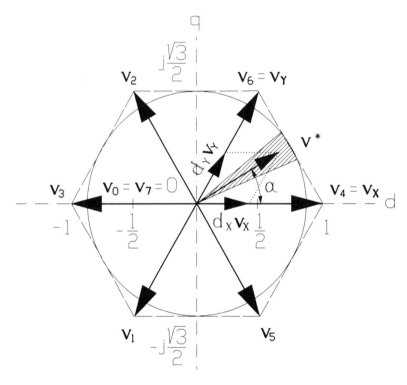

**Figure 5.5.** Illustration of the space vector PWM strategy.

The non-zero base vectors divide the cycle into six, 60°-wide sectors. The desired voltage vector, $v^*$, located in a given sector, can be synthesized as a linear combination of the two adjacent base vectors, $v_X$ and $v_{Y'}$, which are framing the sector, and either one of the two zero vectors, i.e.,

$$\boldsymbol{v}^* = d_X\boldsymbol{v_X} + d_Y\boldsymbol{v_Y} + d_Z\boldsymbol{v_Z} \qquad (5.13)$$

where $v_Z$ is the zero vector, while $d_X$, $d_Y$, and $d_Z$ denote the so-called duty ratios of states $X$, $Y$, and $Z$ within the switching interval, respectively. For example, the reference voltage vector, $v^*$, in Figure 5.5 is located within a sector in which $v_X = v_4$ and $v_Y = v_6$, hence it can be produced by an appropriately timed sequence of states 4, 6, and 0 or 7 of the inverter.

The state duty ratio is defined as the ratio of the duration of the state to the duration of the switching interval. Hence,

$$d_X + d_Y + d_Z = 1. \qquad (5.14)$$

It can be shown that under this condition the locus of the maximum available vectors $v^*$ constitutes the hexagonal envelope of the base vectors, shown in Figure 5.5 as a dashed line. To avoid low-order voltage harmonics, resulting from the non-circular shape of the envelope, the locus of the synthesized voltage vectors is, in practice, limited to the circle inscribed in the envelope, also shown in Figure 5.5. In consequence, the maximum available magnitude, $V_{max}$, of the resulting voltage vector is $\sqrt{3}/2\ V_{dc}$. In order to obtain the maximum available r.m.s. value of the fundamental line-to-neutral voltage of the inverter, this value must be divided by $1.5\sqrt{2}$ (see Eq. (1.60)), then giving $V_{dc}/\sqrt{6} \approx 0.408\ V_{dc}$. Hence, the corresponding maximum available r.m.s. line-to-line voltage is $V_{dc}/\sqrt{2} \approx 0.707\ V_{dc}$. Comparing this value with that for the square-wave operating mode, reveals that the PWM mode results in a 9.4% reduction in the d.c. to a.c. voltage conversion ratio. Yet, assuming that the inverter is supplied from a six-pulse rectifier, which infers that a supply d.c. voltage to the inverter equals $3\sqrt{2}/\pi$ of the r.m.s. line-to-line a.c. voltage supplying the rectifier, the maximum overall, a.c. to a.c., voltage conversion ratio of the rectifier-inverter combination is $3/\pi$, i.e., close to unity.

With respect to vector $v^*$ in Figure 5.5, Eq. (5.13) can be written as

$$\boldsymbol{v}^* = MV_{max}e^{j\alpha} = d_X\boldsymbol{v_4} + d_Y\boldsymbol{v_6} + d_Z\boldsymbol{v_Z} \qquad (5.15)$$

where $M$ is the so-called modulation index, or magnitude control ratio, adjustable within the 0 to 1 range, and $\alpha$ denotes the angular position of vector $v^*$ inside the sector, i.e., the angular distance between vectors $v^*$ and $v_X$. As seen in Figure 5.5, $v_X = v_4 = 1 + j0$ p.u., $v_Y = v_6 = 1/2 + j\sqrt{3}/2$ p.u, and $v_Z$ (either $v_0$ or $v_7$) is zero. Also, as already indicated, $V_{max} = \sqrt{3}/2$

p.u. Substituting these values in Eq. (5.15) and resolving the vectors into $dq$ components, the following equations are obtained:

$$\frac{\sqrt{3}}{2} M\cos(\alpha) = d_X + \frac{1}{2} d_Y \qquad (5.16)$$

$$\frac{\sqrt{3}}{2} M\sin(\alpha) = \frac{\sqrt{3}}{2} d_Y. \qquad (5.17)$$

Solving Eqs. (5.16) and (5.17) for $d_X$ and $d_Y$ gives

$$d_X = M\sin(60°-\alpha) \qquad (5.18)$$

$$d_Y = M\sin(\alpha) \qquad (5.19)$$

while, from Eq. (5.14),

$$d_Z = 1 - d_X - d_Y. \qquad (5.20)$$

The same equations apply to the other sectors, since the $dq$ reference frame, which has here no specific orientation in the physical space, can be aligned with any base vector.

The simple algebraic formulas (5.18) through (5.20) allow duty ratios of the consecutive logic states of an inverter to be computed in real time, in the microprocessor-based modulator. In a similar manner to that for the square-wave operating mode, the inverter operates as a sequential, timed ring counter. Due to the freedom of choice of the zero-vector states, various state sequences can be enforced in a given sector. Particularly efficient operation of the inverter is obtained when the state sequences in consecutive switching intervals are

$$|X - Y - Z|Z - Y - X|\ldots \qquad (5.21)$$

where $Z = 0$ in sectors $v_6 - v_2$, $v_3 - v_1$, and $v_5 - v_4$, and $Z = 7$ in the remaining sectors.

The PWM mode is characterized by the switching frequency being much higher than the output frequency. In general, the higher the ratio of switching frequency to output frequency, the higher is the quality of the output currents so obtained. However, high switching frequencies result in proportionately high switching losses in the inverter switches. Therefore, the switching frequency must represent a reasonable trade-off between the quality of current waveforms and the efficiency of the inverter. In most practical PWM inverters, switching frequencies are limited to a few kHz.

Example switching signals and output voltages for a VSI operating in the previously described PWM mode are shown in Figure 5.6 for $M = 0.75$. The width of the switching intervals is $20°$, hence there are $N = 18$ intervals per cycle of output voltage. In spite of this, the number of pulses per phase, $N_p$, is only 7, i.e., $N_p = N/3 + 1$ per phase. Since $N_p$ represents the ratio of switching frequency to output frequency, a low value of $N_p$ results in low switching losses. In contrast, many PWM techniques produce as many as $N$ pulses per phase, with the output currents of only slightly higher quality, but with much higher switching losses in the inverter.

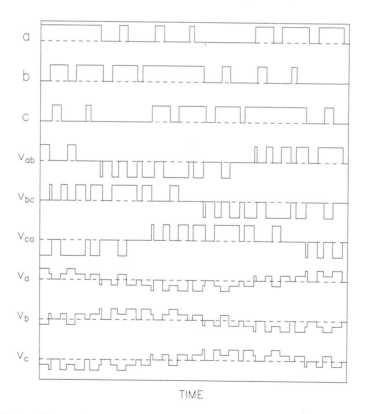

**Figure 5.6.** Example switching signals and output voltages for a VSI in the PWM operating mode.

The harmonic spectrum of the line-to-line voltage of a PWM VSI, for $M = 1$, and $N = 72$ ($N_p = 25$), is shown in Figure 5.7. In comparison with the corresponding spectrum in the square-wave operating mode (see Figure 5.4), the fundamental is somewhat lower and the higher harmonics are significantly different. The lowest harmonic with an amplitude greater than 0.001 p.u. is the 17th, and distinct clusters of prominent harmonics appear around the multiples of $N/2$, i.e., 36, 72, etc. Clearly, such a spectrum is favorable from the point od view of the quality of the output currents, and a further increase in the switching frequency, if feasible, could lead to the production of almost harmonic-free, sinusoidal currents.

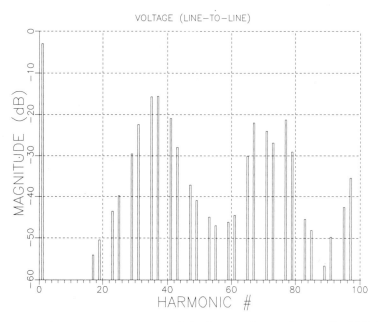

**Figure 5.7.** Harmonic spectrum of the line-to-line voltage of a VSI in the PWM operating mode.

The previously described method belongs to a common class of deterministic, regular-sampling PWM techniques, i.e., those in which the switching patterns, under steady-state operating conditions of the inverter, are replicated from cycle to cycle of the output frequency, and a constant, integer number of equal switching intervals is maintained. However, it is worth noting that a different concept, that of the so-called random pulse modulation (RPWM), has been recently attracting increasing attention. RPWM uses random changes of the widths of the individual switching intervals or positions of the pulses in consecutive intervals. This results in the corresponding power spectra of the inverter acquiring a continuous part and in a reduction of the discrete (harmonic) part of the spectra. In consequence, the acoustic and vibration effects in the associated drive systems are significantly improved. The annoying tonal noise, typical in motors fed from deterministically-modulated VSIs, is replaced by soothing "static," and resonant vibration is unlikely to appear in such a drive system due to the absence of periodic harmonic torques in the motor.

## 5.3 Current Control in Voltage Source Inverters

Since the output currents of an inverter depend on the load, feedforward current control is not feasible, and a feedback from current sensors is required. Again, as in the case of voltage-controlled VSIs, a number of different control techniques exist. Therefore, only the simplest technique, based on the so-called "hysteretic", or "bang-bang", controllers is presented in this treatment, to illustrate the general principle of operation of current-controlled VSIs.

The block diagram of a current-controlled VSI is shown in Figure 5.8. The output currents, $i_a$, $i_b$, and $i_c$, of the inverter are sensed and compared with the reference current signals, $i_a^*$, $i_b^*$, and $i_c^*$. Current error signals, $\Delta i_a$, $\Delta i_b$, and $\Delta i_c$, are then applied to the hysteretic current controllers, which generate switching signals, $a$, $b$, and $c$, for the inverter switches.

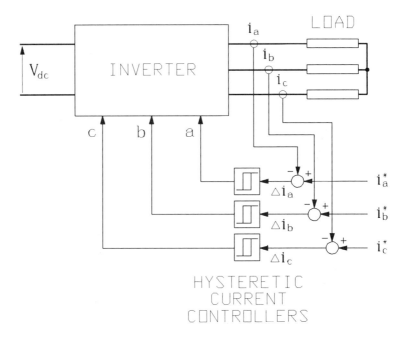

**Figure 5.8.** Block diagram of a current-controlled VSI.

INVERTERS

The input-output characteristic of the phase-A hysteretic current controller is shown in Figure 5.9. The width of the hysteresis loop, denoted by $h$, represents the tolerance bandwidth for the controlled current. If the current error, $\Delta i_a$, is greater than $h/2$, i.e., current $i_a$ is unacceptably lower than the reference current, $i_a^*$, the corresponding line-to-neutral voltage, $v_a$, must be increased. As seen from Eq. (5.10), this voltage is most strongly affected by the switching variable $a$, hence it is this variable that is regulated by the controller, and is set to a 1 in the described situation. Conversely, an error less than $-h/2$ results in $a = 0$, in order to decrease the voltage and current in question. No action is taken by the controller when current $i_a$ stays within the tolerance band. The other two controllers operate in a similar manner.

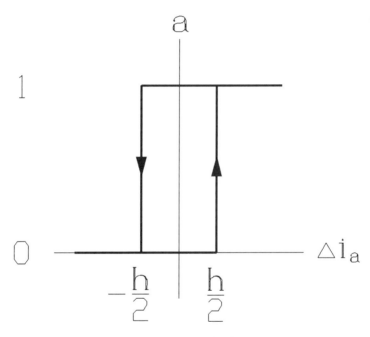

**Figure 5.9.** Input-output characteristic of a hysteretic current controller.

The width, $h$, of the tolerance band affects the switching frequency of the inverter. The narrower is the band, the more frequent switching takes place and the higher quality of the currents. This is illustrated in Figures 5.10 and 5.11, depicting the switching variables, line-to-neutral voltages, and currents for an inverter supplying a resistive-plus-inductive load at values of $h$ equal to 10% and 5% of the amplitude of the reference current, respectively. In practice, the tolerance bandwidth should be set to a value that represents an optimal tradeoff between the quality of the currents and the efficiency of the inverter.

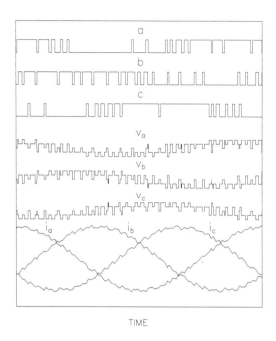

**Figure 5.10.** Current-controlled VSI: 10% tolerance bandwith.

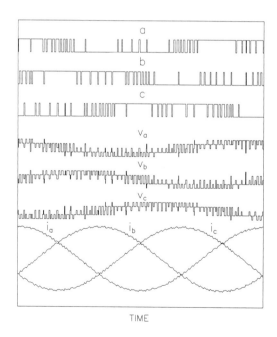

**Figure 5.11.** Current-controlled VSI: 5% tolerance bandwith.

# INVERTERS

## 5.4 Current Source Inverter

The circuit diagram of a three-phase CSI, shown in Figure 5.12, differs from that of a VSI by the absence of freewheeling diodes. Moreover, the d.c. link consists only of an inductor, while the rectifier supplying the d.c. voltage is usually of an SCR-based, controlled type. The inductor maintains the d.c. supply current, $I_{dc}$, at an approximately constant level, hence the supply system of the inverter can be considered as a current source.

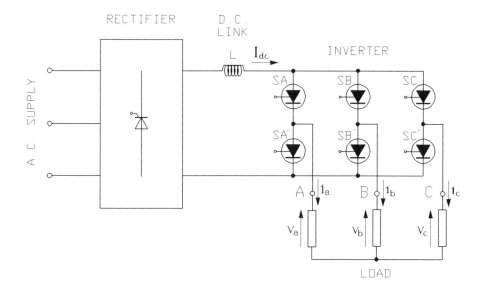

**Figure 5.12.** Circuit diagram of a three-phase CSI.

For the circuit of Figure 5.12, if, for instance, the inverter is in state 6, i.e., $a = 1$, $b = 1$, and $c = 0$, then the d.c. supply current flows through the closed switches SA, SB, and SC'. Assuming a balanced load, the output currents are $i_a = i_b = I_{dc}/2$ and $i_c = -I_{dc}$. Analogous analysis of the other states leads to the following equations:

$$i_a = \frac{I_{dc}}{2}(2a-b-c) \qquad (5.22)$$

$$i_b = \frac{I_{dc}}{2}(2b-c-a) \qquad (5.23)$$

$$i_c = \frac{I_{dc}}{2}(2c-a-b). \qquad (5.24)$$

Eqs. (5.22) through (5.24) are similar to Eqs. (5.10) through (5.12) for a VSI. Therefore, if the 4 - 6 - 2 - 3 - 1 - 5 ... sequence of logic states is imposed on a CSI, stepped current waveforms are produced, as shown in Figure 5.13. The sinusoidal waveforms denoted by $i_{a1}$, $i_{b1}$, and $i_{c1}$ represent the fundamental components of the output currents. The r.m.s. value of these components, equal to $3/\sqrt{2}\pi \, I_{dc} \approx 0.675 \, I_{dc}$ (not drawn to scale in Figure 5.13). Again, as in a VSI, control of the magnitude of the output currents must be performed in the rectifier supplying the inverter. Frequency control is realized by adjusting the duration of the individual states of the inverter, while the phase of the currents can be controlled, to a resolution of 60°, by state changes. This last feature is illustrated in Figure 5.14. Following a portion of the regular sequence, 4 - 6 - 2 - 3 ..., state 3 is changed to state 4, instead to state 1, and the regular sequence is then resumed. The result is that the phase of the fundamental currents has been instantaneously changed by -240° (from 210° to -30°).

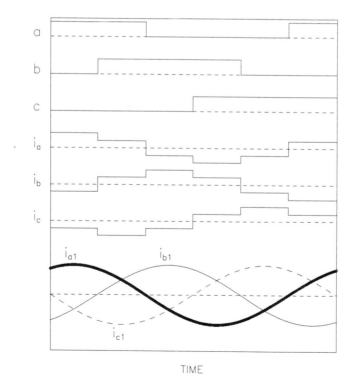

**Figure 5.13.** Switching signals and output currents of a CSI.

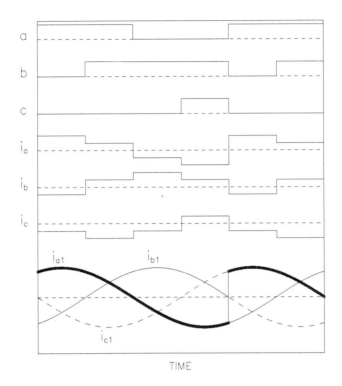

**Figure 5.14.** Phase change of the output currents in a CSI.

The feasibility of frequency, magnitude, and phase control of the output currents makes the CSI suitable for torque control in drive systems employing a.c. motors, in both scalar and vector control versions. The low switching frequency allows the use of relatively slow, high-power semiconductor switches, such as SCRs and GTOs. Therefore, CSIs, based on these switches, are typically employed in high-power drives.

Recently, PWM CSIs came under intensive development. Such an inverter is shown in Figure 5.15. Comparing it with the classic CSI of Figure 5.12, two different features can be discerned. Firstly, the inverter is supplied from an uncontrolled, instead of controlled, rectifier. This implies that the magnitude control of the output currents has been shifted from the rectifier to the inverter. Secondly, the inverter is equipped with three capacitors connected between the output terminals.

Operation of a PWM CSI resembles that of a VSI in the PWM mode. A number of current pulses are generated in each phase of the inverter within a single cycle of output frequency. The switching pattern is such that the fundamental of the pulsed current waveforms has the desired frequency and magnitude. Most of the high-frequency component of these pulses closes through the capacitors, not reaching the load, and waveforms of currents flowing through the load are similar to these produced in a PWM VSI, i.e., rippled sinusoids. Both open-loop and closed-loop control of the output currents is feasible, the latter particularly suitable for vector controlled drive systems.

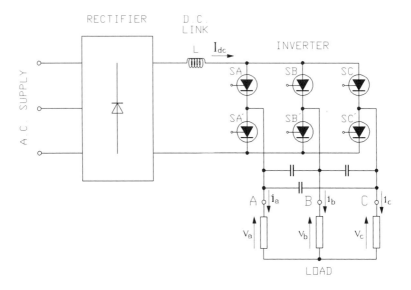

**Figure 5.15.** Circuit diagram of a PWM CSI.

One of the major advantages of CSIs is an inherent cabability to handle both positive and negative power flow in the drive system. The d.c. link current is always positive, but the mean voltage of the link depends on the power supplied to the inverter. Hence, negative power flow is associated with a negative d.c. component of the link voltage. Such voltage polarity can easily be produced in a controlled rectifier or a chopper. Typically, SCR-based rectifiers are used. As is known from the theory of such rectifiers, if the Thevenin's equivalent e.m.f. of the load is negative and the firing angle of the SCRs is greater than 90°, the rectifier operates in the so called line-commutated inverter mode, with the power flowing from the load to the a.c. supply.

## 5.5 Examples and Simulations

The following examples and simulations illustrate the operation of VSIs

# INVERTERS

and CSIs, and the effects of the non-ideal voltage and current waveforms produced by these inverters on the controlled variables in drive systems with induction motors.

## Example 5.1.

The state sequence 4 - 6 - 2 - 3 - 1 - 5 ... of a VSI operating in the square-wave mode results in a positive phase sequence, $A$ - $B$ - $C$, of the output voltages (see Figure 5.3). Determine the state sequence for a negative phase sequence, $A$ - $C$ - $B$, and calculate the duration of each state for an output frequency of 40 Hz. Also, find the d.c. supply voltage of the inverter such that the fundamental line-to-line output voltage is 220 V.

### Solution

The change in phase sequence is obtained by interchanging the switching signals for phases B and C of the inverter. The state sequence for the positive phase sequence can be written in binary notation as

$abc$ = 100-110-010-011-001-101...

Interchanging the last two digits of each state gives

$abc$ = 100-101-001-011-010-110...

which, in decimal notation, represents the **4 - 5 - 1 - 3 - 2 - 6** ... sequence of logic states of the inverter. The duration of each state should be one sixth of the period of the output frequency, i.e., 1/40/6 = 0.00417 sec = **4.17 msec**.

Since the fundamental line-to-line voltage of the inverter equals $\sqrt{6}/\pi$ of the d.c. supply voltage, $V_{dc}$ (see Section 5.2), then $V_{dc} = \pi \times 220/\sqrt{6} =$ **282 V**.

## Example 5.2.

A voltage-controlled VSI employs the space vector PWM technique described in Section 5.2. The inverter is supplied from a 230 V, three-phase line, via a six-pulse rectifier and d.c. link. The number of switching intervals per cycle of output voltage is 72. The inverter is required to produce a fundamental line-to-line voltage of 180 V with a frequency of 100 Hz. Determine the sequence and durations of the inverter states within the

19th switching interval and calculate the switching frequency. Neglect voltage drops in the rectifier, d.c. link, and inverter.

## Solution

A six-pulse rectifier supplied from a three-phase line with the line-to-line r.m.s. voltage of $V_{L-L}$ produces an average output d.c. voltage, $V_{dc}$, equal

$$V_{dc} = \frac{3\sqrt{2}}{\pi} V_{L-L}.$$

In this case, $V_{dc} = 3\sqrt{2} \times 230/\pi$ V = 311 V, and the maximum available r.m.s. value of line-to-line voltage for the inverter is $V_{dc}/\sqrt{2} = 311/\sqrt{2}$ V = 220 V (see Section 5.2). Hence, the modulation index required to obtain an output voltage of 180 V is $M = 180/220 = 0.818$.

Since the 360° cycle of output voltage is divided into $N = 72$ switching intervals, then each interval, in angular terms, is 5° wide. As a result, the 19th switching interval covers the 90° to 95° interval of the output voltage, and the center angle of this interval is 92.5°. This means that the interval in question is located in the sector framed by voltage vectors $v_X = v_6$ and $v_Y = v_2$ of the inverter (see Figure 5.5). The angular position of the reference voltage vector, $v^*$, with respect to vector $v_X$ is $\alpha = 92.5° - 60° = 32.5°$. The zero vector, $v_Z$, used in this sector is $v_0$ (see Eq. 5.21), and the duty ratios of the corresponding states, calculated from Eqs. (5.18) through (5.20), are

$$d_6 = 0.818 sin(60°-32.5°) = 0.378$$

$$d_2 = 0.818 sin(32.5°) = 0.440$$

$$d_0 = 1-0.378-0.440 = 0.182.$$

Since each sector is divided into 72/6 = 12 switching intervals, the 19th interval is the seventh interval in the $v_6$ - $v_2$ sector and, according to Eq. (5.21), the state sequence is X-Y-Z, i.e., 6-2-0. The period of the output frequency is 10 msec, hence a switching interval is 10/72 msec = 0.139 msec = 139 μsec long. Consequently, state 6 is maintained for 0.378 × 139 μsec = 53 μsec, state 2 for the next 0.440 × 139 μsec = 61 μsec, and state 0 for the remaining 25 μsec of the interval duration.

The number of pulses per phase is

$$N_p = \frac{N}{3}+1 = \frac{72}{3}+1 = 25.$$

Hence, the switching frequency of the inverter switches is $25 \times 100$ Hz = 2.5 kHz.

### Example 5.3.

A CSI is used in the vector control system for an induction motor. At a certain instant of time, the reference amplitude, $I_{max}^*$, and phase angle, $\omega t$, signals for the phase-A output current are 50 A and 130°, respectively. Determine the next logic state of the inverter and the required d.c. supply current.

*Solution*

It is assumed that the fundamental phase-A current, $i_{a1}$, is expressed as a cosinusoid, i.e., as

$$i_{a1} = I_{max}\cos(\omega t).$$

As seen in Figure 5.13, the state sequence 4 - 6 - 2 - 3 - 1 - 5 produces a full cycle of this current, starting at a phase angle, $\omega t$, of -30° and ending at a phase angle of 330°. This means that the transition to state 4 takes place at -30°, that to state 6 at 30°, etc. Thus, the reference phase angle of 130° is closest to the phase angle of transition from state 2 to state 3, i.e., $\omega t = 150°$, and it is **state 3** that should be imposed on the inverter.

The r.m.s. value of the fundamental output currents is $3/\sqrt{2\pi}\, I_{dc}$ (see Section 5.4). Hence, the d.c. current required for the given amplitude of the currents is

$$I_{dc} = \frac{\pi}{3} \times 50 = \mathbf{52.4\ A}.$$

### Simulation 5.1. Voltage Source Inverter - Square-Wave Operating Mode

To illustrate the effect of a VSI operating in the square-wave mode on an induction motor, a simulation was carried out with the example motor running under steady-state, rated conditions.

Stator voltage and current waveforms are shown in Figure 5.16. The stepped voltage waveforms result in a heavily distorted current. As a consequence of this, the developed torque, shown in Figure 5.17, acquires an oscillatory component, called a periodic harmonic torque. The frequency of this torque is six times that of the stator currents. In this simulation, the harmonic torque has little effect on the speed of the motor, also shown in Figure 5.17, since the mechanical inertia of the drive system provides effective damping. However, at low values of the supply frequency, this damping may be inadequate to prevent speed oscillations. Also, the harmonic torque may cause accelerated mechanical wear of joints and bearings in the system.

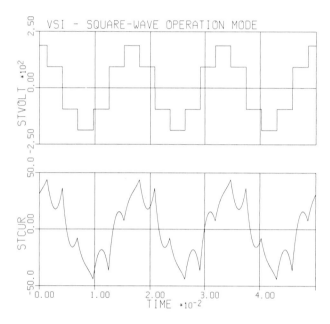

**Figure 5.16.** Stator voltage and current of a VSI-fed motor: square-wave operating mode.

The rotor current waveform and magnitude of the rotor flux are shown in Figure 5.18. It is interesting to note that in spite of the fact that the waveform of the rotor current is similarly distorted to that of the stator current, the rotor flux is maintained at an almost steady level.

The square wave operation of the inverter results not only in the production of the undesired harmonic torque, but also in increased losses in the motor, caused by the strong harmonic content of the stator and rotor currents. A partial, albeit expensive, countermeasure to these problems can be provided by addition of extra inductors between the inverter and the motor, in order to improve the quality of the currents.

# INVERTERS

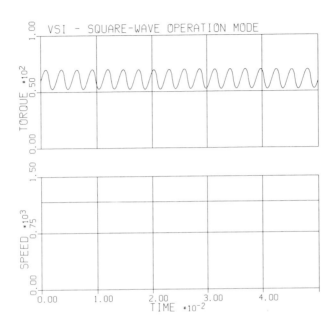

**Figure 5.17.** Torque and speed of a VSI-fed motor: square-wave operating mode.

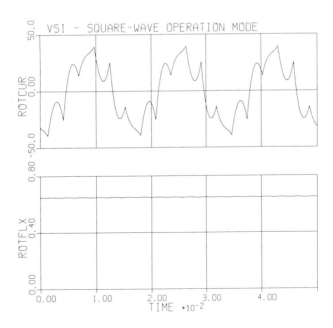

**Figure 5.18.** Rotor current and flux of a VSI-fed motor: square-wave operating mode.

## Simulation 5.2. Voltage Source Inverter - PWM Operating Mode

The same drive system as that for Simulation 5.1 is considered here, but with the inverter operating in the PWM mode, employing the space vector technique described in Section 5.2. A cycle of output voltage is divided into 30 switching intervals. Waveforms of the stator voltage and current are shown in Figure 5.19, the torque and speed of the motor in Figure 5.20, and the rotor current and flux in Figure 5.21.

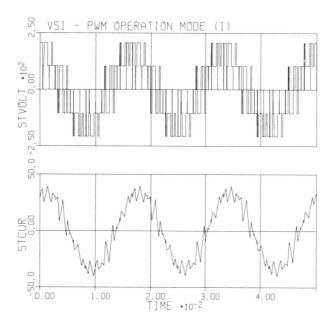

**Figure 5.19.** Stator voltage and current waveforms of a VSI-fed motor: PWM operating mode, 30 switching intervals per cycle.

The pulsating stator voltage produces a current, with superimposed ripple of a significantly higher frequency, and lower magnitude than in the case of the square-wave operating mode (see Figure 5.15). The harmonic torque also has increased high-frequency content, although the 360 Hz component is still dominant. The rotor current waveform is similar to that of the stator current and, again, the rotor flux is practically constant.

The same quantities are shown in Figures 5.22 through 5.24, but with the number of switching intervals per cycle increased to 60. An improvement in the current waveforms and the resulting reduction of the harmonic torque (albeit at the expense of higher switching losses in the inverter) are easily observable.

INVERTERS 151

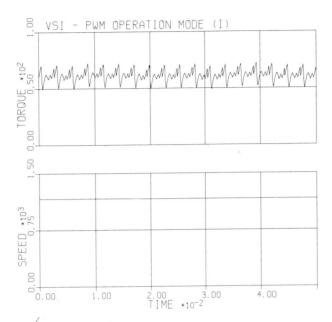

**Figure 5.20.** Torque and speed of a VSI-fed motor: PWM operating mode, 30 switching intervals per cycle.

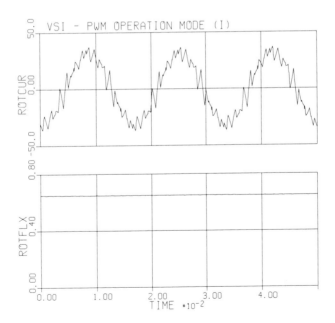

**Figure 5.21.** Rotor current and flux of a VSI-fed motor: PWM operating mode, 30 switching intervals per cycle.

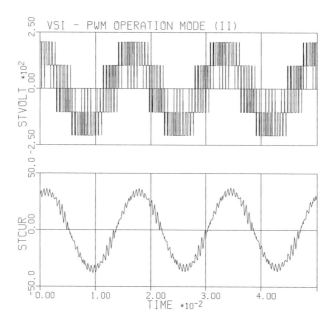

**Figure 5.22.** Stator voltage and current waveforms of a VSI-fed motor: PWM operating mode, 60 switching intervals per cycle.

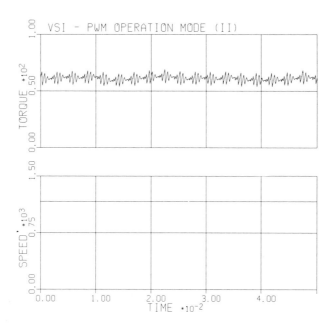

**Figure 5.23.** Torque and speed of a VSI-fed motor: PWM operation mode, 60 switching intervals per cycle.

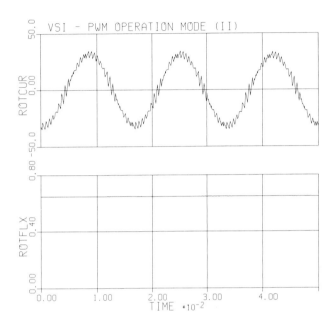

**Figure 5.24.** Rotor current and flux of a VSI-fed motor: PWM operating mode, 60 switching intervals per cycle.

**Simulation 5.3. Vector Control System with Current-Controlled Voltage Source Inverter**

To illustrate the effect of an inverter on the operation of a vector controlled motor, the same drive system as in Simulation 4.2 is modelled, but with a current-controlled VSI in place of the previously assumed ideal source of stator currents. The tolerance bandwidth, $h$, is 5 A.

Variations of the torque and speed are shown in Figure 5.25. It can be seen that the demanded time profiles of the torque and speed are followed with somewhat lower accuracy than in the idealized system, and a harmonic torque is superimposed on the piecewise constant, programmed torque.

The stator current and supply current of the inverter are shown in Figure 5.26 for the time interval within which the transition from small positive to large negative torque takes place. The stator current waveform is seen to be of high quality, and a rapid change in phase angle, peculiar to vector control, can also be observed. The supply current is complex, with a high-frequency a.c. component, and a d.c component which changes in proportion to the motor torque.

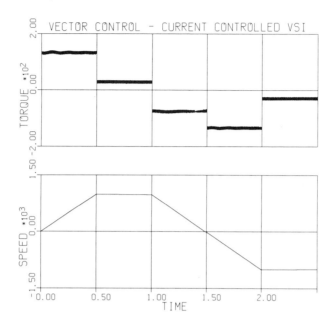

**Figure 5.25.** Torque and speed variations for the motor in a vector controlled drive system with a current-controlled VSI.

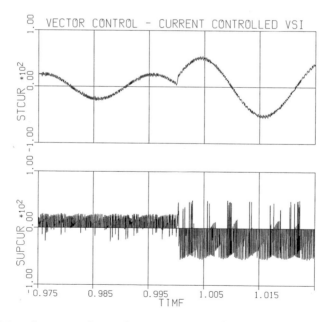

**Figure 5.26.** Stator and supply current waveforms during the transition from positive to negative torque.

## Simulation 5.4. Current Source Inverter

This simulation illustrates steady-state operation of the example motor, when supplied from a CSI and driving a rated load at rated speed. This condition requires stator currents with an r.m.s. value of 23.8 A (see Example 1.1). Consequently, the d.c. supply current of the inverter must be $\sqrt{2\pi}/3 \times 23.8$ A = 35.2 A (see Section 5.4). The frequency of the stator currents is 60 Hz, therefore, each inverter state lasts 1/360 sec $\approx$ 2.78 msec.

The stator current and voltage waveforms are shown in Figure 5.27. The voltage waveform is practically sinusoidal, but with a spike at each state transition of the inverter. These spikes result from the voltage drop, proportional to the derivative of the stator current, across the stator leakage inductance. Since idealized, discontinuous current waveforms were assumed, the amplitudes of the spikes are theoretically infinite. Clearly, this situation could not be allowed in practice. Therefore, motors with the lowest possible flux leakage in the stator are selected for such applications, while the commutation interval of the inverter, i.e., the interval during which a given branch takes over conduction from another branch, is extended so that the rate of change of current is not excessive. The result is that the voltage spikes are limited to the extent that they do not endanger the inverter switches and/or the motor insulation.

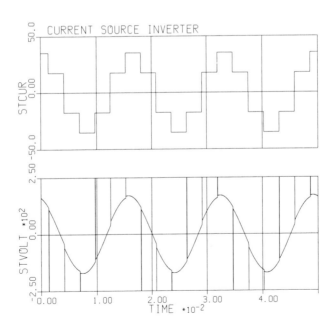

**Figure 5.27.** Stator current and voltage waveforms of a CSI-fed induction motor.

The torque waveform, shown in Figure 5.28, displays a distinct, 360 Hz periodic component generated by the stepped current waveforms. However, the mechanical inertia of the drive system provides such an effective damping, that the motor speed, also shown in this figure, is practically unaffected. This feature of the CSI-fed motor resembles that of the motor driven by a VSI in the square-wave operation mode, illustrated in Figure 5.17.

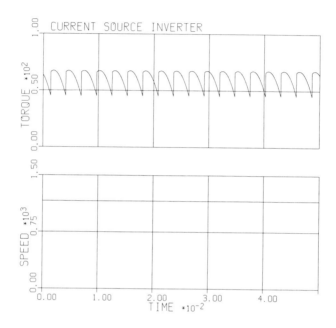

**Figure 5.28.** Torque and speed of a CSI-fed induction motor.

The waveform of the rotor current and magnitude of the rotor flux waveforms are shown in Figure 5.29. The rotor current is broadly similar to the stator current. The flux is practically constant, although some minor ripple can be discerned.

The presence of a harmonic torque constitutes a major disadvantage of the classic, six-step CSIs, since it produces extra noise and vibration in the supplied drive system. This is the reason for the ongoing research on the PWM CSIs, mentioned in Section 5.4. However, although very promising, these inverters have not yet been used to any extent in commercial practice.

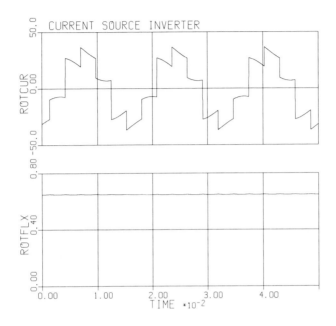

**Figure 5.29.** Rotor current and flux of a CSI-fed induction motor.

## Chapter 6

## REVIEW OF VECTOR CONTROL SYSTEMS

Both the direct and indirect field orientation schemes presented in Chapter 4 have specific advantages and disadvantages. The direct scheme requires use of flux sensors installed inside the motor and spoiling the ruggedness of the machine. Instead of the delicate Hall sensors, the so-called search coils can be used which allow the airgap flux measurement to be performed by integrating the induced voltage. These, however, are useless at low speeds, which result in an excessively long integration time. The indirect approach to field orientation is based on sensing the position of the rotor, again a frail arrangement, and, most of all, it is very sensitive to motor parameters. Incorrect knowledge of these leads to significant deterioration of the operation of the drive system. Therefore, a tremendous amount of research has been conducted to develop viable vector control systems for induction motors in commercial applications. This effort is far from being concluded, and only time will show which of the large variety of solutions proposed every year in scores of papers will survive the test of practical implementation. In effect, only a general review of the existing vector control systems can be attempted in this book.

Ongoing research focuses primarily on the three following approaches:

(1) Alternative alignment of the rotating excitation frame. Although the classic orientation of this frame along the rotor flux vector results in the automatic decoupling of the torque and field controls, the other two fluxes, i.e., the stator flux and airgap flux, can also be used with certain advantages over the rotor flux orientation scheme.

(2) Elimination of direct sensors of the motor quantities and replacing them with indirect observers, particularly with respect to flux and speed estimation.

(3) Adaptive schemes, involving on-line identification of motor parameters and tuning of the control system.

These concepts, as well as the use of CSIs in vector control of induction motors, are reviewed in this chapter. Finally, basic issues of the speed and position control using field-oriented induction motors are presented.

## 6.1 Systems with Stator Flux Orientation

As shown in Chapters 3 and 4, the fundamental device of the FOP is the elimination of the negative term in the cross-product part of the torque equation of an induction motor. By setting $\lambda_{QR}^e$ to zero, Eq. (3.5) has been reduced to Eq. (3.11) and the developed torque has been made proportional to the product of the mutually orthogonal components, $\lambda_{DR}^e$ and $i_{QS}^e$, of the vectors of rotor flux and stator current. Then, according to Eq. (3.12), $\lambda_{DR}^e$ depends only on $i_{DS}^e$. Consequently, the torque and flux of a motor can be controlled independently by adjusting the torque-producing, $i_{QS}^e$, and flux-producing, $i_{DS}^e$, components of the stator current vector $\mathbf{i}_S^e$.

It can easily be shown that the torque developed in an induction motor can be expressed in terms of a cross-product of vector $\mathbf{i}_S^e$ and any one of the three flux vectors $\lambda_S^e$, $\lambda_M^e$, and $\lambda_R^e$, the latter being used in Eq. (3.11). This section decribes systems with field orientation employing the stator flux vector. The rationale behind this choice of the reference vector is that the estimation of magnitude and phase of the stator flux vector is considered to be easier than that of the rotor flux vector.

Rearranging Eq. (1.86), the rotor current vector can be expressed as

$$\mathbf{1}_R^e = \frac{1}{L_m}(\boldsymbol{\lambda}_S^e - L_s \mathbf{1}_S^e) \qquad (6.1)$$

which, when substituted in Eq. (1.50), yields

$$T = \frac{P}{3}(i_{QS}^e \lambda_{DS}^e - i_{DS}^e \lambda_{QS}^e) \ . \qquad (6.2)$$

If the $D$-axis of the excitation reference frame is aligned with the stator flux vector $\lambda_s^e$, then $\lambda_{QS}^e = 0$, and

$$T = k_T i_{QS}^e \lambda_{DS}^e \qquad (6.3)$$

where $k_T = P/3$. Clearly, Eq. (6.3) is analogous to Eq. (3.11) of a motor in a vector control system with rotor flux orientation.

Substituting Eq. (6.1) in Eq. (1.87) gives

$$\boldsymbol{\lambda}_R^e = \frac{L_r}{L_m} (\boldsymbol{\lambda}_s^e - \sigma L_s \boldsymbol{i}_s^e) \tag{6.4}$$

where $\sigma$ denotes the so-called total leakage factor, defined as

$$\sigma \equiv 1 - \frac{L_m^2}{L_s L_r}. \tag{6.5}$$

If $\lambda_{QS}^e = 0$, then,

$$\boldsymbol{i}_R^e = \frac{1}{L_m} [\lambda_{DS}^e - L_s (i_{DS}^e + j i_{QS}^e)] \tag{6.6}$$

$$\boldsymbol{\lambda}_R^e = \frac{L_r}{L_m} [\lambda_{DS}^e - \sigma L_s (i_{DS}^e + j i_{QS}^e)]. \tag{6.7}$$

Substituting Eqs. (6.6) and (6.7) in Eq. (3.4), yields

$$\lambda_{DS}^e = \frac{1 + \sigma \tau_r (p + j\omega_r)}{1 + \tau_r (p + j\omega_r)} L_s (i_{DS}^e + j i_{QS}^e). \tag{6.8}$$

Clearly, the $\lambda_{DS}^e$ component of the stator flux vector in a field-oriented motor may not have the imaginary component implied by Eq. (6.8). Also, the magnitude, $\lambda_s = \lambda_{DS}^e$, of this vector should be independent of the $i_{QS}^{e*}$ component of the stator current vector. Therefore, the reference signal, $i_{DS}^{e*}$, of the flux-producing current must be made dynamically dependent on the reference signal, $i_{QS}^{e*}$, of the torque-producing current, in such a way that $i_{QS}^{e*}$ has no effect on $\lambda_{DS}^e$ and that $\lambda_{DS}^e$ is always a real number.

The required decoupling equations can be obtained by separating Eq. (6.8) into the real and imaginary parts, which are

$$\frac{\lambda_{DS}^e}{L_s} = \frac{1+\sigma\tau_r p}{1+\tau_r p} i_{DS}^e - \frac{\sigma\tau_r\omega_r}{1+\tau_r p} i_{QS}^e \qquad (6.9)$$

and

$$\frac{\lambda_{DS}^e}{L_s} = \sigma i_{DS}^e + (1+\sigma\tau_r p) i_{QS}^e. \qquad (6.10)$$

Solving Eq. (6.9) for $i_{DS}^e$ and Eq. (6.10) for $\omega_r$, gives

$$i_{DS}^{e*} = \frac{(p+\frac{1}{\tau_r})\frac{\lambda_s^*}{\sigma L_s} + \omega_r^* i_{QS}^{e*}}{p + \frac{1}{\sigma\tau_r}} \qquad (6.11)$$

$$\omega_r^* = \frac{p + \frac{1}{\sigma\tau_r}}{\frac{\lambda_s^*}{\sigma L_s} - i_{DS}^{e*}} i_{QS}^{e*} \qquad (6.12)$$

where $\lambda_s^*$ is the reference magnitude of the stator flux (independent of the reference frame) and $\omega_r^*$ denotes the reference slip speed of the motor. The decoupling system, which converts the reference flux, $\lambda_s^*$, and torque, $T^*$, signals into reference current signals, $i_{DS}^{e*}$ and $i_{QS}^{e*}$, is shown in Figure 6.1.

Apart from the SOAR conditions (see Section 2.6), the classic rotor flux orientation schemes do not impose any theoretical limits on the torque and flux commands. Such limits, however, exist in the systems with stator flux orientation. The steady-state solution for the slip speed command, $\omega_r^*$, can be determined from Eqs. (6.11) and (6.12). Setting $p$ to zero and eliminating the $i_{DS}^{e*}$ variable, the following quadratic equation is obtained:

$$\omega_r^{*2} - \frac{1-\sigma}{\sigma^2 L_s \tau_r} \frac{\lambda_s^*}{i_{QS}^{e*}} \omega_r^* + \frac{1}{(\sigma\tau_r)^2} = 0. \qquad (6.13)$$

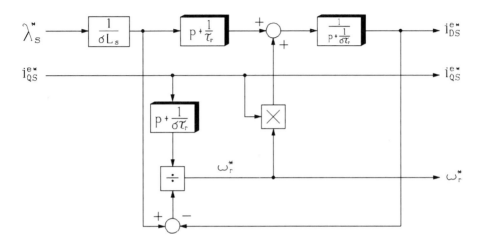

**Figure 6.1.** Decoupling system for stator flux orientation.

For the solution of this equation to be a real number, the discriminant may not be negative. This is only satisfied if

$$(i_{QS}^{e*})^2 \leq [\frac{\lambda_s^*}{2L_s}(\frac{1}{\sigma}-1)]^2 . \qquad (6.14)$$

From Eq. (6.3),

$$i_{QS}^{e*} = \frac{3}{P}\frac{T^*}{\lambda_s^*} , \qquad (6.15)$$

Substituting Eq. (6.15) in Eq. (6.14), and rearranging, gives

$$\frac{|T^*|}{\lambda_s^{*2}} \leq \frac{P}{6L_s}(\frac{1}{\sigma}-1) . \qquad (6.16)$$

Eq. (6.16) implies that it is the leakage factor, σ, that imposes the "torque per flux squared" limitation in stator flux orientation schemes.

As demonstrated in Chapter 4, two general approaches can be used to orient the *D*-axis of the excitation frame along the selected flux vector. In the direct field orientation scheme, the position of the flux vector is directly sensed using a flux sensor or observer. The indirect field orientation method employs the mathematical model of the motor to compute the flux vector position from the rotor position and stator currents. Both these techniques, illustrated in Sections 4.3 and 4.4 for the rotor flux orientation, can be extended on systems with the stator flux orientation.

When controllers for closed-loop regulation of the torque and stator flux of the motor are added to the decoupling system, a reference current system is created, which provides the $i_{DS}^{e*}$ and $i_{QS}^{e*}$ signals in response to the reference flux and torque commands. This system, used in the direct stator flux orientation scheme, is shown in Figure 6.2. The complete vector control system is depicted in Figure 6.3.

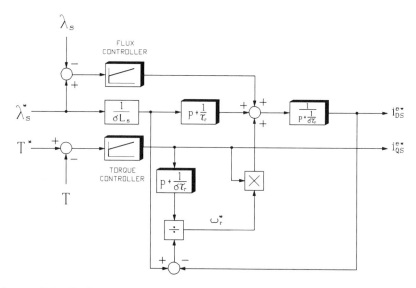

**Figure 6.2.** Reference current system for direct stator flux orientation.

In a manner similar to that involving the rotor flux vector in the system with direct rotor flux orientation (see Section 4.3), the magnitude, $\lambda_s$, and phase, $\Theta_s$, of the stator flux vector $\lambda_s^s$ is derived by the stator flux calculator from sensors of the airgap flux, $\lambda_m^s$, and stator current, $i_s^s$. Relation

$$\lambda_s^s = \lambda_m^s + L_{ls} i_s^s \tag{6.17}$$

# REVIEW OF VECTOR CONTROL SYSTEMS

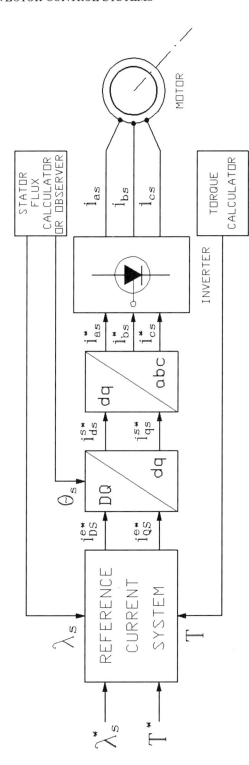

**Figure 6.3.** Vector control system for an induction motor with direct stator flux orientation.

is utilized. Hall sensors, search coils, or taps on stator winding are used to measure the airgap flux vector.

As an alternative to the stator flux calculator, an observer can be employed. In that case, vector $\lambda_s^s$ is determined as

$$\lambda_s^s = \int v_s^s - R_s i_s^s dt \qquad (6.18)$$

i.e., by measuring stator voltages and currents. This solution has the advantage of simplicity, since no sensors inside the motor are required. It is also quite accurate as, of all electric parameters of the equivalent circuit of an induction motor, the stator resistance, $R_s$, is the easiest to evaluate.

The torque calculator is based on the same principle as that in the direct rotor flux orientation scheme and can be disposed of should an open-loop torque control be chosen.

The decoupling network can provide the reference slip speed signal, $\omega_r^*$, for an indirect stator flux orientation. The reference current and slip speed system for this scheme is shown in Figure 6.4, and the complete vector control system with indirect stator flux orientation is illustrated in Figure 6.5.

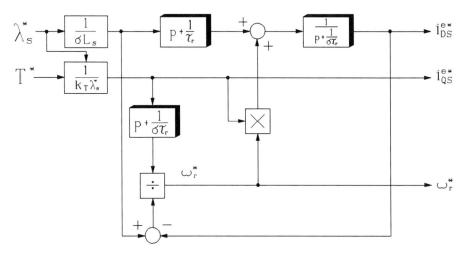

**Figure 6.4.** Reference current and slip speed system for indirect stator flux orientation.

# REVIEW OF VECTOR CONTROL SYSTEMS

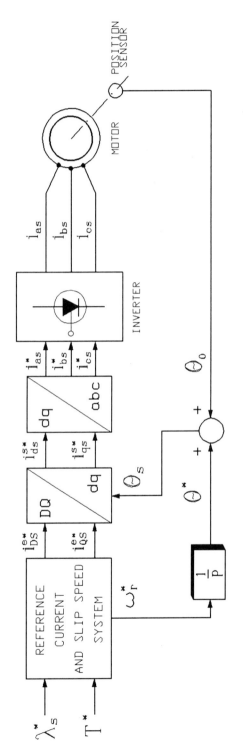

**Figure 6.5.** Vector control system for an induction motor with indirect stator flux orientation.

It can be seen that the decoupling system constitutes the only essential difference between vector control systems with stator flux orientation and their counterparts with rotor flux orientation. No decoupling is required if the rotor flux vector is used to orient the excitation reference frame. This is the reason why the systems with rotor flux orientation have historically been developed first and, consequently, have often been termed "classic." It turns out, however, that the increased complexity of systems with stator flux orientation pays off in terms of easier determination of coordinates of the reference flux vector and the enhanced robustness of these systems, i.e., reduced dependency on precise knowledge of motor parameters.

## 6.2 Systems with Airgap Flux Orientation

In the described vector control systems for induction motors with direct rotor or stator flux orientation, it was the airgap (mutual) flux that was sensed in order to determine the vector of either rotor or stator flux. Therefore, it is logical to use the airgap flux vector, $\lambda_m^s$, for alignment with the $D$-axis of the rotating excitation frame.

The stator flux differs form the airgap flux by only the leakage in the stator. Hence,

$$\lambda_S^e = \lambda_M^e + L_{ls} i_S^e . \tag{6.19}$$

Resolving $\lambda_S^e$ in Eq. (6.19) into the $DQ$ components and substituting in Eq. (6.2) gives

$$T = \frac{P}{3}(i_{QS}^e \lambda_{DM}^e - i_{DS}^e \lambda_{QM}^e) . \tag{6.20}$$

Substituting Eq. (6.19) in Eq. (1.51) yields

$$i_R^e = \frac{\lambda_M^e}{L_m} - i_S^e \tag{6.21}$$

which, when substituted in Eq. (1.87), gives

# REVIEW OF VECTOR CONTROL SYSTEMS

$$\lambda_R^e = \frac{L_r}{L_m}\lambda_M^e - L_{1r}\mathbf{1}_s^e. \qquad (6.22)$$

If the excitation reference frame is oriented so that the $D$-axis is aligned with the airgap flux vector, $\lambda_m^a$, then $\lambda_{QM}^e = 0$, and

$$T = k_T i_{QS}^e \lambda_{DM}^e \qquad (6.23)$$

where $k_T = P/3$,

$$\mathbf{1}_R^e = \frac{\lambda_{DM}^e}{L_m} - i_{DS}^e - j i_{QS}^e \qquad (6.24)$$

$$\lambda_R^e = \frac{L_r}{L_m}\lambda_{DM}^e - L_{1r}(i_{DS}^e + j i_{QS}^e). \qquad (6.25)$$

Substituting Eqs. (6.24) and (6.25), and $v_R^e = 0$, in Eq. (1.85) yields

$$\lambda_{DM}^e = \frac{1+\sigma_r\tau_r(p+j\omega_r)}{1+\tau_r(p+j\omega_r)} L_m(i_{DS}^e + j i_{QS}^e). \qquad (6.26)$$

where $\sigma_r$ is the rotor leakage factor, defined as

$$\sigma_r \equiv \frac{L_{1r}}{L_r}. \qquad (6.27)$$

Again, as in the case of stator flux orientation schemes, a decoupling system must be used for independent control of the flux and torque of the motor and for elimination of the imaginary part of $\lambda_{DM}^e$ in Eq. (6.26).

The real and imaginary parts of Eq. (6.26) can be written as

$$\frac{\lambda_{DM}^e}{L_m} = \frac{1+\sigma_r\tau_r p}{1+\tau_r p} i_{DS}^e - \frac{\sigma_r\tau_r\omega_r}{1+\tau_r p} i_{QS}^e \qquad (6.28)$$

and

$$\frac{\lambda_{DM}^{e}}{L_{m}} = \sigma_{r} i_{DS}^{e} + (1+\sigma_{r}\tau_{r}p) i_{QS}^{e}. \qquad (6.29)$$

Equations of the decoupling system, obtained by solving Eq. (6.28) for $i_{DS}^{e}$ and Eq. (6.29) for $\omega_r$, are

$$i_{DS}^{e*} = \frac{(p+\frac{1}{\tau_{r}})\frac{\lambda_{m}^{*}}{\sigma_{r}L_{m}} + \omega_{r}^{*} i_{QS}^{e*}}{p + \frac{1}{\sigma_{r}\tau_{r}}} \qquad (6.30)$$

$$\omega_{r}^{*} = \frac{p + \frac{1}{\sigma_{r}\tau_{r}}}{\frac{\lambda_{m}^{*}}{\sigma_{r}L_{m}} - i_{DS}^{e*}} i_{QS}^{e*} \qquad (6.31)$$

where $\lambda_{m}^{*}$ denotes the reference magnitude of the airgap flux.

The "torque per flux squared" limitation is

$$\frac{|T^{*}|}{\lambda_{m}^{*2}} \leq \frac{P}{6 L_{m}} (\frac{1}{\sigma_{r}} - 1). \qquad (6.32)$$

Since $\sigma_r < \sigma$, condition (6.32) is less restrictive than that given by Eq. (6.16) for systems with stator flux orientation.

Comparison of Eqs. (6.11), (6.12), and (6.16) with Eqs. (6.30), (6.31), and (6.32) reveals a close analogy between systems with stator flux orientation and airgap flux orientation. Replacing $\lambda_{s}^{*}$ with $\lambda_{m}^{*}$, $L_s$ with $L_m$, and $\sigma$ with $\sigma_r$ in schemes with stator flux orientation, corresponding schemes with airgap flux orientation are obtained.

The decoupling system, current reference system, and complete vector control system for an induction motor with direct airgap flux orientation are shown in Figures 6.6 through 6.8, respectively. The magnitude, $\lambda_m$, and phase, $\Theta_m$, of the airgap flux vector are either directly sensed or estimated by an observer, from vectors of the stator voltage and current, as

REVIEW OF VECTOR CONTROL SYSTEMS    171

$$\lambda_m^s = \int v_s^s - R_s \mathbf{1}_s^s dt - L_{ls} \mathbf{1}_s^s. \qquad (6.33)$$

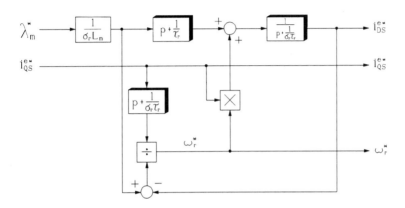

**Figure 6.6.** Decoupling system for airgap flux orientation.

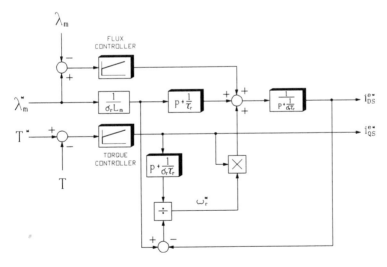

**Figure 6.7.** Reference current system for direct airgap flux orientation.

Figure 6.9 illustrates the reference current and slip speed system for indirect airgap flux orientation. Complete vector control system with this orientation is shown in Figure 6.10.

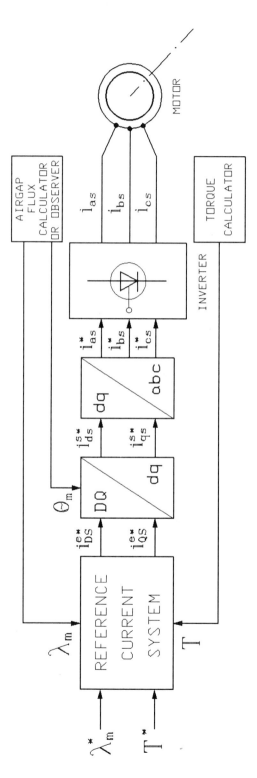

**Figure 6.8.** Vector control system for an induction motor with direct airgap flux orientation.

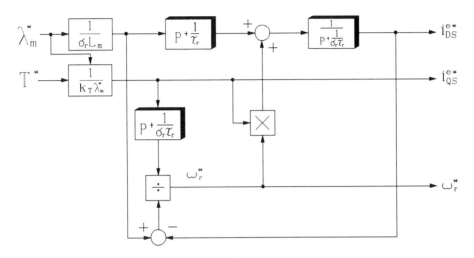

**Figure 6.9.** Reference current and slip-speed system for indirect airgap flux orientation.

The similarities between systems with stator and airgap flux orientation extend to systems with rotor flux orientation. Indeed, if $\lambda_{DM}^e$ in Eq. (6.28) is replaced with $\lambda_{DR}^e$, and $\sigma_r$ with a zero, then a relation equivalent to Eq. (3.12) is obtained. This has given rise to a concept of the so-called universal field orientation (UFO). In a microprocessor controlled UFO system, versatile software allows free choice of the reference flux vector and of the direct or indirect orientation scheme. Two major advantages result from this flexibility:

(1) A single type of vector controller can be produced and easily matched to a given motor and employed sensors or observers. If, for example, the motor is equipped with Hall sensors of the airgap flux, then the best solution is the direct airgap flux orientation scheme. However, for a standard motor, the stator flux observer and, consequently, the stator flux orientation, are more suitable. This versatility of the controller reduces the development and manufacturing expenses.

(2) The operating mode of the system can be adapted to the operating conditions of the motor. For instance, at low speeds, the indirect orientation mode is preferable since flux sensors or observers which utilize an induced voltage perform poorly at low frequencies. On the other hand, at higher speeds of the motor, the direct orientation mode represents a better choice, being less dependent on motor parameters.

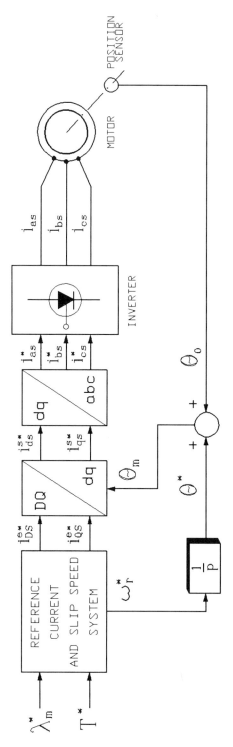

**Figure 6.10.** Vector control system for an induction motor with indirect airgap flux orientation.

## 6.3 Systems with Current Source Inverters

CSIs are often used in high-power vector controlled drives with induction motors. The employed thyristor switches, SCRs or GTOs, are relatively slow and, therefore, more suitable for the square-wave operating mode than for the high-frequency switching required in current controlled VSIs. As explained in Section 5.4, CSI fed from controlled rectifiers have the advantageous capability of two-directional power flow. Reliability of the expensive inverter is also an important consideration. Common for all types of inverters is the problem of the so-called shot-through which occurs when one switch in a given branch is turned on before the other switch has completely turned off. In VSIs, such an occurence causes shorting of the d.c. link capacitance and the resultant dangerous overcurrent. Similar shorting of the current source supplying a CSI is clearly less hazardous.

As explained in Chapter 5, the current control in CSIs differs from that in VSIs. In the latter inverters, output currents follow the reference signals which represent the required sequential instantaneous values of the currents. The current control is performed in the inverter alone by the closed-loop governed switching of the inverter switches. CSIs, instead, react to the magnitude and phase commands, the magnitude of the currents being adjusted in the controlled rectifier supplying the d.c. link, and only the phase being enforced in the inverter in the open-loop mode. Therefore, vector control systems with CSIs differ from these with VSIs with regard to the reference signals provided to the inverter.

A simplified block diagram of control system of a CSI in a vector controlled drive system is shown in Figure 6.11. The reference stator current signals, $i_{ds}^{s*}$ and $i_{qs}^{s*}$, from the field orientation part of the system are converted to reference signals, $I_{dc}^*$ and $\varphi_s^*$, of the d.c. link current and phase of the stator current vector, respectively. The d.c. link current is controlled in the rectifier, in a closed loop with a PI-type controller, while the reference phase signal enforces an appropriate state of the inverter.

Practical systems are more complicated, since the problem of sluggish response of the both converters must be dealt with. Also, the d.c. link current is influenced by the back e.m.f. of the motor. Various solutions have been proposed, usually involving dynamic lead-type compensators and estimators of the back e.m.f. These, however, exceed the scope of this book.

It should be mentioned that drive systems with CSIs perform poorly at low speeds because of strong torque pulsations resulting from the stepped waveforms of stator currents. Significant improvement in this respect is expected from introduction of CSIs with sinusoidal currents, mentioned in Section 5.4 and now being under development.

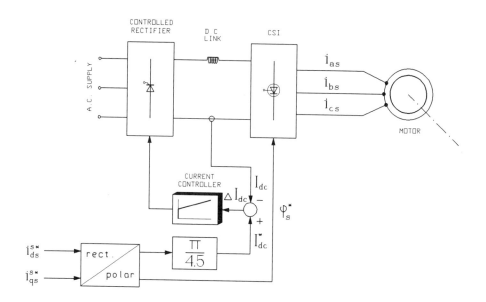

**Figure 6.11.** Control system of a CSI in a vector control system.

## 6.4 Observers for Vector Control Systems

Sensors of flux and speed or position of the rotor spoil the ruggedness of drive systems with induction motors. Therefore, there is tendency to replace them with observers which convert the stator voltage and current signals into the required information concerning other variables of the motor. Voltage and current transducers have traditionally been installed in electric drive systems for monitoring purposes; hence, the input signals for the observers are easily available. If precise estimation of a given flux vector is required, a speed sensor is used as a compromise solution between a fully sensorless motor and one with direct flux sensors.

Observers used in vector control systems are either of an open-loop or closed-loop type. In both cases, motor equations are used to estimate a given variable from measured values of other variables. Accuracy of the open-loop observer strongly depends on the assumed values of parameters of the motor. Therefore, in the closed-loop arrangement, two independent estimators (sub-observers) are used, each based on an equation or set of

REVIEW OF VECTOR CONTROL SYSTEMS 177

equations different from that employed by the other estimator. An adaptation mechanism is provided to converge the outputs of the two estimators. In effect, parameter sensitivity of the closed-loop observer is significantly lower than that of each of the two estimators.

The following example of observers for estimation of the rotor speed, $\omega_o$, illustrates the difference between the open-loop and closed-loop concepts. Eq. (1.26) allows expressing the vectors of stator flux, $\lambda_s^s$, and rotor current, $i_r^s$, in terms of vectors $\lambda_r^s$ and $i_s^s$ of rotor flux and stator current, as

$$\lambda_s^s = \frac{L_m}{L_r}\lambda_r^s + \sigma L_s i_s^s \qquad (6.34)$$

$$i_r^s = \frac{1}{L_r}(\lambda_r^s - L_m i_s^s). \qquad (6.35)$$

Substituting Eqs. (6.34) and (6.35) in Eqs. (1.24) and (1.25), and setting $v_r^s$ in Eq. (1.25) to zero, gives, after simple rearrangements,

$$p\lambda_r^s = \frac{L_r}{L_m}[v_s^s - (R_s + \sigma L_s p)i_s^s] \qquad (6.36)$$

$$p\lambda_r^s = (j\omega_o - \frac{1}{\tau_r})\lambda_r^s + \frac{L_m}{\tau_r}i_s^s. \qquad (6.37)$$

Following the $dq$ resolution, Eqs. (6.36) and (6.37) can be expressed in the matrix form as

$$p\begin{bmatrix}\lambda_{dr}^s \\ \lambda_{qr}^s\end{bmatrix} = \frac{L_r}{L_m}\left(\begin{bmatrix}v_{ds}^s \\ v_{qs}^s\end{bmatrix} - \begin{bmatrix}R_s + \sigma L_s p & 0 \\ 0 & R_s + \sigma L_s p\end{bmatrix}\begin{bmatrix}i_{ds}^s \\ i_{qs}^s\end{bmatrix}\right)$$

$$(6.38)$$

and

$$p\begin{bmatrix}\lambda_{dr}^s\\ \lambda_{qr}^s\end{bmatrix} = \frac{1}{\tau_r}\left(L_m\begin{bmatrix}i_{ds}^s\\ i_{qs}^s\end{bmatrix} - \begin{bmatrix}1 & \omega_o\tau_r\\ -\omega_o\tau_r & 1\end{bmatrix}\begin{bmatrix}\lambda_{dr}^s\\ \lambda_{qr}^s\end{bmatrix}\right).$$

(6.39)

Eqn. (6.38) allows estimation of $\lambda_{dr}^s$ and $\lambda_{qr}^s$. Then, the magnitude, $\lambda_r$, and phase, $\Theta_r$, of the rotor flux vector, $\lambda_r^s$, can be calculated as

$$\lambda_r = \sqrt{(\lambda_{dr}^s)^2 + (\lambda_{qr}^s)^2} \qquad (6.40)$$

$$\Theta_r = \tan^{-1}\left(\frac{\hat{\lambda}_{qr}^s}{\hat{\lambda}_{dr}^s}\right). \qquad (6.41)$$

From Eq. (6.41), the derivative, $p\Theta_r$, of angle $\Theta_r$, is

$$p\Theta_r = \frac{1}{\lambda_r^2}(\lambda_{dr}^s \times p\lambda_{qr}^s - \lambda_{qr}^s \times p\lambda_{dr}^s). \qquad (6.42)$$

Substituting $p\lambda_{dr}^s$ and $p\lambda_{qr}^s$ from Eq. (6.39) in Eq. (6.42) gives

$$p\Theta_r = \omega_o + \frac{L_m}{\tau_r\lambda_r^2}(i_{qs}^s\lambda_{dr}^s - i_{ds}^s\lambda_{qr}^s). \qquad (6.43)$$

Note that the second term in Eq. (6.43) is proportional to the torque developed in the motor (see Eq. (3.3)), as

$$\frac{L_m}{\tau_r\lambda_r^2}(i_{qs}^s\lambda_{dr}^s - i_{ds}^s\lambda_{qr}^s) = \frac{3R_r}{P\lambda_r^2}T. \qquad (6.44)$$

Consequently, from Eqs. (6.43) and (6.44),

REVIEW OF VECTOR CONTROL SYSTEMS 179

$$\omega_o = p\Theta_r - \frac{3R_r}{P\lambda_r^2}T \qquad (6.45)$$

where $p\Theta_r$ is determined from Eq. (6.42).

A block diagram of the observer is shown in Figure 6.12. It can be seen that besides the estimated speed of the motor, estimates of other important variables, such as the motor torque or magnitude and angle of the rotor flux vector, can also be obtained.

Since accurate knowledge of motor parameters is required for satisfactory performance of the open-loop observer described, alternative solutions, employing the adaptive control theory, have been developed. A closed-loop speed observer, based on the Model Reference Adaptive System (MRAS) technique, constitutes a typical example of such an approach.

The basic concept of the closed-loop speed observer is illustrated in Figure 6.13. Two flux estimators, FE-1 and FE-2, independently evaluate the components of the rotor flux vector in the stator reference frame. FE-1 utilizes the stator equation (6.38), while FE-2 uses the rotor equation (6.39). Since the stator equation does not include the estimated quantity, $\omega_o$, FE-1 may be considered as a reference model, and FE-2 as an adjustable model. State difference, $\varepsilon$, of the two models is applied to a linear controller which produces the estimate of motor speed for the adjustable model.

It has been proven that if a proportional-plus-integral (PI) controller is used, i.e.,

$$\omega_o = K_P\varepsilon + K_I\int_0^t \varepsilon\, dt \qquad (6.46)$$

then the observer, which is a nonlinear feedback system, is stable when

$$\varepsilon = \lambda_{qr}^s \lambda_{dr}^{s\prime} - \lambda_{dr}^s \lambda_{qr}^{s\prime} \qquad (6.47)$$

where $\lambda_{dr}^s$ and $\lambda_{qr}^s$ are estimates of the $dq$ components of the rotor flux vector obtained from FE-1, while $\lambda_{dr}^{s\prime}$ and $\lambda_{qr}^{s\prime}$ are these generated by FE-2.

A detailed diagram of the closed-loop speed observer is shown in Figure 6.14. The observer performs better and is somewhat simpler than the open-

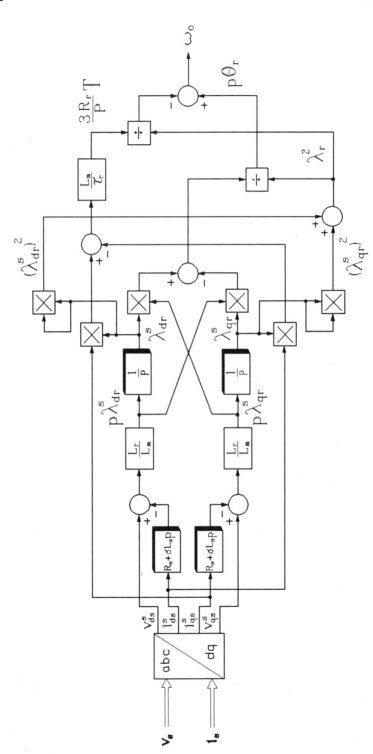

**Figure 6.12.** Open-loop speed observer.

loop observer in Figure 6.12. Note that the value of rotor resistance, $R_r$, has been conveniently incorporated into the controller constants $K_P$ and $K_I$.

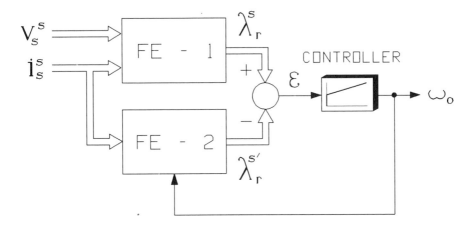

**Figure 6.13.** Illustration of the concept of closed-loop speed observer.

For a truly high-performance vector control system, the degree of accuracy of the described observers may turn out as insufficient. In such a case, a closed-loop rotor flux observer can be used, with an additional input from the sensor of speed of the rotor.

A general block diagram of the observer is shown in Figure 6.15. The observer consists of three flux estimators, FE-1, FE-2, and FE-3, and a linear controller. Two independent estimates of the rotor flux vector, $\lambda_r^s$ and $\lambda_r^{s\prime}$ are obtained from FE-2 and FE-3, respectively. Based on Eq. (6.34), FE-2 calculates $\lambda_r^s$ from the stator flux vector, $\lambda_s^s$, evaluated in FE-1, and stator current vector, $i_s^s$, as

$$\pmb{\lambda}_r^s = \frac{L_r}{L_m}(\pmb{\lambda}_s^s - \sigma L_s \pmb{i}_s^s) \qquad (6.48)$$

while FE-3 utilizes Eq. (6.37). The state difference, $\varepsilon$, of the two estimators is converted in the controller into a signal $v_\varepsilon^s$ which is fed back to FE-1. There, the estimate of the stator flux vector is computed from Eq. (1.53), augmented with the $v_\varepsilon^s$ term, as

$$p\pmb{\lambda}_s^s = \pmb{v}_s^s - R_s \pmb{i}_s^s + \pmb{v}_\varepsilon^s. \qquad (6.49)$$

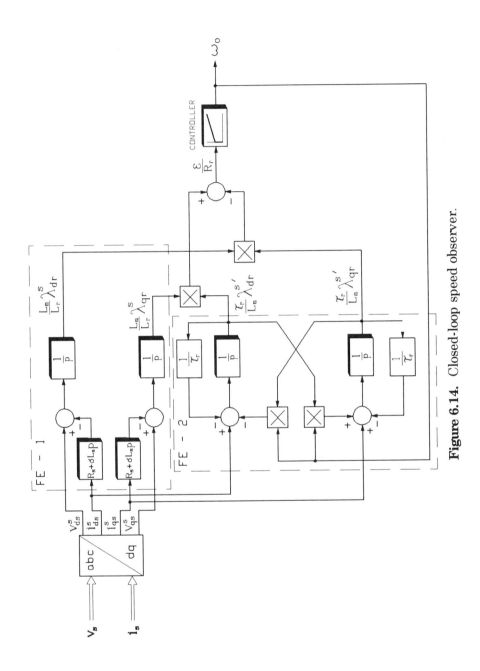

**Figure 6.14.** Closed-loop speed observer.

# REVIEW OF VECTOR CONTROL SYSTEMS

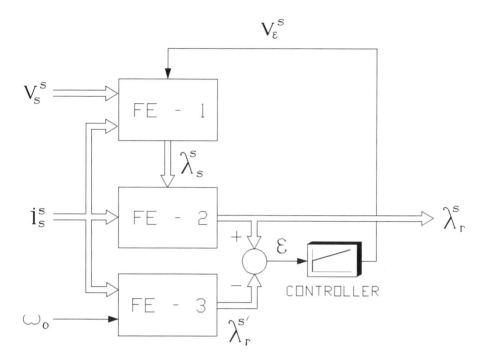

**Figure 6.15.** Illustration of the concept of closed-loop rotor flux observer.

A full diagram of the closed loop rotor flux observer is shown in Figure 6.16. Note that the individual signals represent vector quantities, i.e., complex numbers, without the resolution into their dq components. Such format of block diagrams can be encountered in literature on control of a.c. machines.

It must be stressed that the presented observers constitute only selected examples, provided for illustration of the basic concepts of estimation of motor variables. A large variety of observers has been proposed in the literature on vector control of a.c. motors. Interestingly, some of them are based on the physical phenomena occuring in real machines but neglected in the idealized mathematical model used in this book. For instance, saturation of the magnetic circuit of an induction motor causes induction of a third-harmonic, zero-sequence voltage component in stator windings. This component, obtained by addition of stator phase voltages, allows determination of the spatial position of the airgap flux vector, $\Theta_m$ (see Figure 6.10). Also, in most machines the airgap flux density includes a high-frequency component, generated by the stator and rotor slots and related to the rotor speed. Therefore, it can be used for indirect, yet accurate, estimation of $\omega_o$.

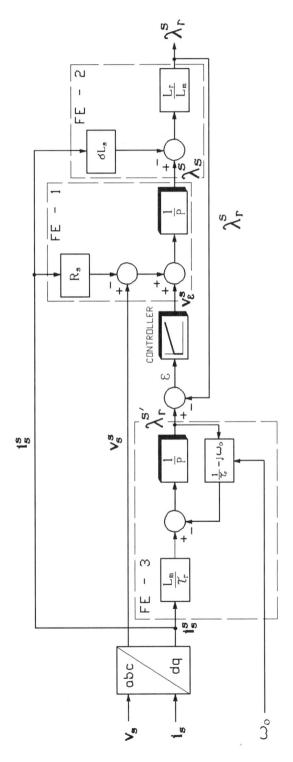

**Figure 6.16.** Closed-loop rotor flux observer.

## 6.5 Adaptive Schemes

The dependence of performance of a vector control system on accurate knowledge of parameters of the controlled motor has been repeatedly stressed. Resistances of the windings increase with temperature and the rotor resistance is additionally affected by the slip frequency due to the skin effect. Inductances are also varying, because of saturation of the magnetic circuit of the motor. Therefore, even laboratory tests performed on a machine prior to the commissioning of the drive system provide information that is often insufficient for the required level of performance of the system, especially if the indirect flux orientation method is used.

Adaptive schemes, integrated with the basic structure of vector control systems, allow continuous or periodical identification of selected motor parameters and/or on-line tuning of the system. For instance, vector control systems for induction motors display particular sensitivity to variations of the rotor time constant, $\tau_r$. This is mainly dependent on temperature. Therefore, attempts were made to evaluate $\tau_r$ on a basis of measurements of stator temperature or, indirectly, by employing a thermal model of the motor. Clearly, such techniques could produce rough estimates only. Rapid progress in the areas of integrated circuits and signal processing now allows implementation of more sophisticated and accurate methods of parameter identification.

Just as for observers, the two basic approaches used for the on-line identification of motor parameters can be classified as open-loop and closed-loop. Open-loop techniques utilize the inverter as a source of a specific voltage or current input signal to the stator. Analysis of the motor response, i.e., stator voltage waveform if a current excitation was employed or stator current if a voltage signal was applied, allows determination of the parameters of the equivalent circuit of the motor. Closed-loop schemes are based on the MRAS theory, having already been illustrated in the preceding section. The actual state of a motor is compared with that of an adaptive reference model, and the state difference is used to adjust the model parameters until satisfactory convergence is obtained.

A simple method of determining of the rotor time constant, used in the so-called self-commissioning drive systems, utilizes the periods of time when the motor is at a standstill for at least several seconds. The inverter is then employed as a source of d.c. current supplied to the motor in order to build up a stationary magnetic field. The motor represents a three-phase transformer with shorted secondary, rotor windings. Because of the d.c. form of the primary, stator current, no current is induced in the rotor. When, in the next step, the inverter is turned off, the stator windings are open and the electromagnetic energy stored in the field is dissipated in the rotor resistance. This means that currents appear in the rotor windings, dying out in time because of exhaustion of the electromagnetic energy. These time-varying currents induce certain voltages in the stator windings.

As d.c. quantities, current and voltage vectors can be made to have only the $d$-axis component by setting the field-producing d.c. stator currents in accordance to the diagram shown in Figure 1.2, i.e., $i_{bs} = i_{cs} = -i_{as}/2$. When the excitation has ceased, then, with $\bar{i}_s^s = 0$ and $\omega_o = 0$, the matrix equation (1.27) can be rewritten as

$$v_{ds}^s = L_m \frac{di_{dr}^s}{dt} \tag{6.50}$$

$$R_r i_{dr}^s + L_r \frac{di_{dr}^s}{dt} = 0. \tag{6.51}$$

Eq. (6.51) can be rearranged to

$$\frac{di_{dr}^s}{dt} = -\frac{1}{\tau_r} i_{dr}^s \tag{6.52}$$

and solved, giving

$$i_{dr}^s(t) = i_{dr}^s(0) e^{-\frac{t}{\tau_r}} \tag{6.53}$$

where $i_{dr}^s(0)$ is the initial value of the rotor current at the instant when the stator currents cease to flow. Substituting Eq. (6.53) in Eq. (6.52) and, next, Eq. (6.52) in Eq. (6.50) yields

$$v_{ds}^s(t) = -\frac{L_m}{\tau_r} i_{dr}^s(0) e^{-\frac{t}{\tau_r}}. \tag{6.54}$$

From Eq. (6.54),

$$\frac{v_{ds}^s(\tau_r)}{v_{ds}^s(0)} = \frac{1}{e} = 0.368 \tag{6.55}$$

which means that the rotor time constant can be determined as the amount of time during which the induced stator voltage has decreased from its initial value to 36.8% of this value.

The method described is not applicable for updating the values of motor parameters in continuous-motion drive systems, and not highly accurate, because of the somewhat idealized picture of the actual physical phenomena involved. Recently, continuous identification schemes based on the MRAS theory have been gaining popularity. A block diagram of an adaptive estimator of motor parameters is shown in Figure 6.17.

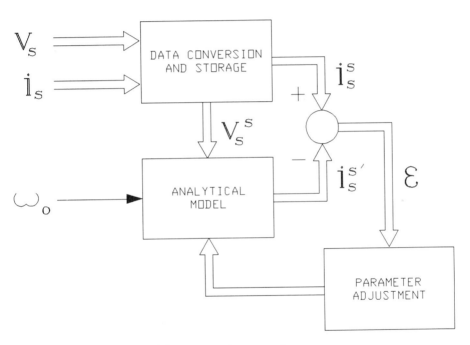

**Figure 6.17.** Adaptive estimator of motor parameters.

Stator voltages and currents are sensed, converted into the $dq$ components of the respective stator-frame vectors, and stored as functions of time. The voltage signals, $v_{ds}^s$ and $v_{qs}^s$, and the speed signal, $\omega_o$, from a rotor speed sensor, are applied to the analytical model of the motor, whose initial parameter settings have been obtained from the catalog data or laboratory tests of the motor. Since each component of the voltage vector generated by the voltage source inverter has only 5 possible values (see Figure 5.2), then a closed-form solution of the dynamic equation (1.35) of the motor can be predetermined for each of them. Solutions representing the $dq$ components of the stator current vector as functions of time allow calculation of the estimated trajectory of this vector. The adaptation error, $\varepsilon$, between the actual trajectory, $i_s^a$, and the estimated trajectory, $i_s^{a\prime}$, is minimized in an iterative process. The values of motor parameters in the analytical model are adjusted using the method of the maximum gradient of the error. The model parameters assume their final values when a preset allowable value of the error is reached, and they are then considered to be sufficiently close to the actual parameters of the motor.

Parameter estimation schemes significantly increase the complexity of the control system. Therefore, alternative approaches to adaptive tuning of vector control systems have been developed. As an example, a simple technique using the torque feedback to adjust the reference slip speed signal, $\omega_r^*$, in a system with indirect rotor flux orientation, described in Section 4.4, is presented. Clearly, this signal is crucial for high performance of the drive system. However, as indicated by Eq. (4.18) and illustrated in Figure 6.18 (which is a fragment of Figure 4.8), $\omega_r^*$ is inversely proportional to the rotor time constant, $\tau_r$. This constant, in practice, is the most unpredictable parameter of an induction motor.

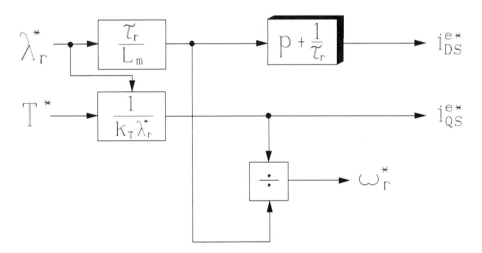

**Figure 6.18.** Open-loop control of the reference slip speed.

The adaptive method of control of the reference slip speed is illustrated in Figure 6.19. The $\omega_r^*$ signal is corrected by subtracting from it an adaptation signal $\Delta\omega_r^*$, generated by a PI-type controller which is activated by the adaptation error, $\varepsilon_2$. The adaptation error is obtained by multiplying a control error, $\varepsilon_1$, by the reference torque-producing current, $i_{QS}^*$, where the control error represents a difference between absolute values of the reference torque, $T^*$, and the estimated (as difficult to be measured directly) torque, $T$, of the motor. In this way, the dependence of $\omega_r^*$ on the rotor time constant is weakened.

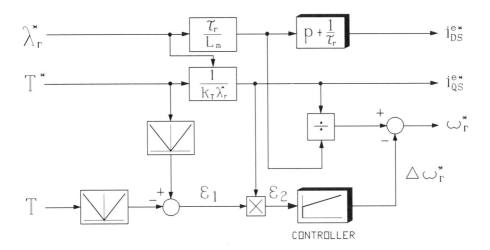

**Figure 6.19.** Adaptive control of the reference slip speed.

The torque estimator, which uses the measured stator voltage and current, is shown in Figure 6.20. The estimator, based on Eq. (6.2), is fairly accurate, particularly at moderate and high values of frequency of the stator currents, when practical variations of the stator resistance have a negligible effect on the stator flux.

## 6.6 Position and Speed Control of Field-Oriented Induction Motors

A vector control system for an induction motor allows decoupling of the torque and flux controls and ensures the optimal torque production in the motor. In practice, vector control systems are seldom self-sufficient, but rather constitute subsystems of larger schemes whose purpose is efficient and accurate control of the position and/or speed of the mechanical load driven by the motor. Direct control of only the developed torque has, simply, limited commercial applications. A notable exception is the manual-control mode of the drive system for an electric vehicle, in which the accelerator pedal provides the reference torque signal. Even there, in the cruise-control mode, it is the speed of the vehicle, and no longer the torque of the motor, that becomes the directly controlled variable.

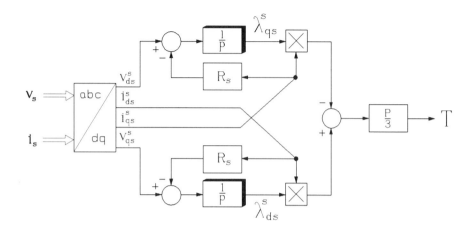

**Figure 6.20.** Torque estimator.

Robustness, i.e., insensitivity to parameter variations, load changes, and external disturbances, constitutes a major requirement for the positioning and adjustable-speed electric drive systems. Based on the theory of servomechanisms, several robust control schemes have, over the years, been developed, usually for a d.c. motor as the motion actuator. These schemes can easily be adapted to field-oriented induction motors, which conform to the dynamic model of a d.c. motor.

Typically, the position control systems include at least two feedback loops: an outer loop for the position, $\Theta_M$, of the shaft of the motor, and an inner loop for the speed, $\omega_M$, of the motor. A speed control system requires at least one feedback loop, for the controlled speed $\omega_M$. If a direct (feedback) flux orientation scheme is employed in the vector controller of the motor, an additional loop for control of the developed torque, $T$, appears in the system as a part of the vector controller. The classic position control scheme, illustrated in Figure 6.21, employs linear, PI or PID (proportional-plus-integral-plus-derivative) controllers in the individual feedback loops. Since settings of the controllers are fixed, this scheme is characterized by relatively low robustness, having been optimally tuned to only a given, single set of parameters of the system.

The corresponding configuration of a speed control system can be obtained directly from Figure 6.21 by removing the outer, position control, loop.

# REVIEW OF VECTOR CONTROL SYSTEMS

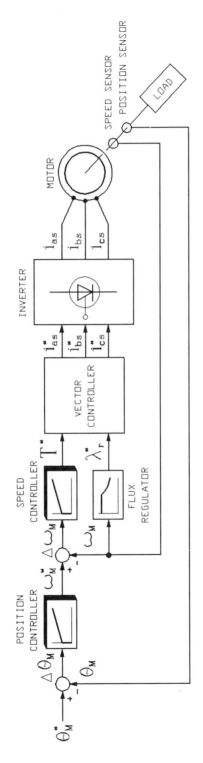

**Figure 6.21.** Position control system for a field-oriented induction motor with linear controllers.

The robustness of a control system can be greatly enhanced by applying the Variable Structure Control (VSC). This recently very popular control method, based on the concept of the so-called sliding mode, has been proven to be very advantageous for vector controlled drive systems.

To explain the basic principles of the VSC, a linear dynamic system will be considered. The state-space representation of the system is

$$\dot{x} = Ax + Bu \qquad (6.56)$$

where $x$ denotes an $n$-dimensional vector of state variables, $u$ is an $m$-dimensional vector of inputs, while $A$ and $B$ are $n \times n$ and $n \times m$ matrices, respectively. In the classic, linear-feedback control systems, such as that in Figure 6.21, the inputs constitute continuous functions of time and the gains of controllers are constant. The VSC systems employ various control methods, the most popular being the relay control and switched-gains control. The relay control results in discontinuous input signals, while the switched-gains control consists in switching the gain of a given controller between two preassigned values. In both the cases, the control vector, $u(x)$, is governed by the sign of an $m$-dimensional, so-called, vector switching function $s(x)$. Components $s_i(x)$, $i = 1, 2, \ldots, m$, of the vector switching function are called scalar switching functions, and are assigned to the individual $m$ inputs to the system. Each scalar switching function describes a linear surface (hyperplane) $s_i(x) = 0$.

Let $x_0$ be the initial state of the system at the initial time $t_0$, $x(t)$ be the state at any time $t$, and $S$ be a switching surface that includes the origin $x = 0$. If, for any $x_0$ in $S$, $x(t)$ is also in $S$ for all $t > t_0$, then $x(t)$ is called a sliding mode (sliding motion) of the system. Moreover, if for every point in $S$ there exist trajectories $x(t)$ reaching it from both sides of $S$, then switching surface $S$ is called a sliding surface. The condition under which the state, $x$, of the system will move toward and eventually reach a sliding surface is called a reaching condition. When the reaching condition is satisfied, the state of the system reaches this surface in a finite time, and stays within a preassigned tolerance band about the sliding surface for all the subsequent time.

Various formulations of the reaching condition have been proposed in the literature, the simplest of which is

$$s_i \frac{ds_i}{dt} < 0 \qquad (6.57)$$

which means that if $s_i$ is negative, i.e., below the sliding surface, then it should be forced to increase to reach the surface and, vice versa, when $s_i$ is positive, i.e., above the surface, then it should be forced to decrease toward the surface.

The relay control law is

$$u_i(\mathbf{x}) = \begin{cases} U_i^+ & \text{if } s_i(\mathbf{x}) > 0 \\ U_i^- & \text{if } s_i(\mathbf{x}) < 0 \end{cases} \qquad (6.58)$$

where $u_i(\mathbf{x})$ denotes control of the $i$-th input, while constants $U_i^+$ and $U_i^-$ represent the positive and negative relay gains, respectively. They may be either fixed or state dependent, and chosen to satisfy the desired reaching condition.

The switched-gains control law can be expressed as

$$u_i(\mathbf{x}) = \sum_{j=1}^{n} g_{ij}(\mathbf{x})(x_j^* - x_j) \qquad (6.59)$$

where

$$g_{ij}(\mathbf{x}) = \begin{cases} G_{ij}^+ & \text{if } s_i(\mathbf{x})(x_j^* - x_j) > 0 \\ G_{ij}^- & \text{if } s_i(\mathbf{x})(x_j^* - x_j) < 0 \end{cases} \qquad (6.60)$$

and $x_j^*$ denotes the desired, reference value of state variable $x_j$. Similarly to the constants $U_i^+$ and $U_i^-$ in the relay control, gains $G_{ij}^+$ and $G_{ij}^-$ may be either fixed or state dependent, and must satisfy the desired reaching condition. VSC systems combining the two control methods described are also feasible, as are systems with the combined, variable-structure and linear-feedback controls.

The VSC is applicable to both linear and non-linear systems and is particularly recommended for systems with model uncertainty. Since the dynamic characteristics of a VSC system depend on the employed sliding surface, any feasible dynamic performance of the system can be enforced by a proper design of the control algorithm. Therefore, VSC systems are counted among the most robust control arrangements.

The position or speed control of an electric drive system constitutes the, so-called, tracking problem, where the state of the system, $x(t)$, is expected to follow a specific reference trajectory, $x^*(t)$. A popular structure of the switching function used for this purpose is

$$s(x) = k(x^* - x) \qquad (6.61)$$

where $k$ is an $m$-dimensional vector of constant coefficients.

A case of the relay control, although not termed as such, has already been presented in Section 5.3, devoted to the current control in VSIs. The hysteretic, relay-type controllers force the output currents of a VSI toward the prescribed reference trajectories which constitute the sliding surfaces. The reaching condition is satisfied if the d.c. supply voltage of the inverter is suffiently high to maintain the currents at the desired levels. Specifically, the maximum available voltage across the load must be higher than the Thevenin's terminal voltage of the load, corresponding to the reference peak current.

The same case illustrates the problem of "chattering", typical for the relay control. The discontinuous inputs cause the state, $x$, of the system to "overshoot" the sliding surface most of the time when $x$ approaches the surface from either side. Currents in a current-controlled VSI do not stay in their sliding surfaces but oscillate within the tolerance band. This is, unfortunately, the price paid for the increased robustness of VSC schemes. In contrast, linear control of the currents would require a linear electronic amplifier in place of a switching inverter and linear controllers replacing the hysteretic controllers. Such a system would, theoretically, be capable of precisely following the reference currents, if the controllers were appropriately tuned to the given operating conditions. However, it would be significantly more sensitive to external disturbances and parameter variations, apart from the fact that linear amplifiers are totally impractical for the application in question.

A variable-structure speed control system with a relay controller is shown in Figure 6.22. The controller, with a hysteretic relay characteristic, causes the reference torque signal, $T^*$, to switch between two preassigned constant values, $T^+$ and $T^-$, typically large but limited to the allowable range of the motor torque. No switching occurs when the speed error, $\Delta\omega_M$, does not exceed the tolerance band determined by the width of the hysteresis loop of the controller.

To reduce the chattering, i.e., speed oscillations around the reference trajectory, $\omega_M^*$, a linear feedback loop can be added in parallel with the relay controller, as shown in Figure 6.23. The linear controller takes over the generation of the reference torque signal when the speed of the motor falls within the tolerance band. No hardware switch is employed in practical systems, the changes in the reference torque signal being accomplished by means of software. However, the switches in the diagrams of Figures 6.22 and 6.23 better illustrate the basic concept of VSC systems and explain the origin of the "variable structure" term.

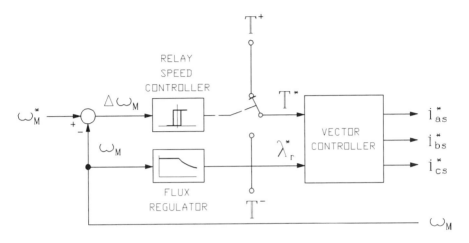

**Figure 6.22.** Variable structure speed control system with a relay controller for a field-oriented induction motor.

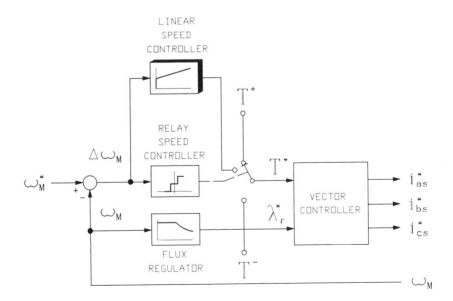

**Figure 6.23.** Variable structure speed control system with a relay controller and a linear controller for a field-oriented induction motor.

In certain applications, such as elevator drives, both the position, $\Theta_M$, and speed, $\omega_M$, of the motor are required to follow certain prescribed trajectories, $\Theta_M^*$ and $\omega_M^*$, respectively. A block diagram of an example system for switched-gains control of position and speed of a field-oriented induction motor is shown in Figure 6.24. The switching function, $s$, represents a weighted sum of the position error, $\Delta\Theta_M$, and speed error, $\Delta\omega_M$, i.e.,

$$s = k_1 \Delta\Theta_M + k_2 \Delta\omega_M \qquad (6.62)$$

where $k_1$ and $k_2$ are constant or state-dependent coefficients. The reference torque signal, $T^*$, is

$$T^* = G_1 \Delta\Theta_M + G_2 \Delta\omega_M \qquad (6.63)$$

where each of the gains $G_1$ and $G_2$ is switched between two values, again constant or state-dependent, $G_1^+$ or $G_1^-$ and $G_2^+$ or $G_2^-$, respectively, by a switching controller realizing Eq. (6.60).

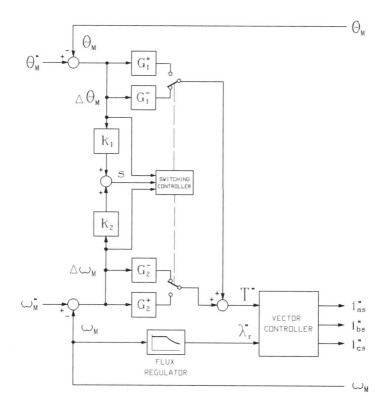

**Figure 6.24.** Switched-gains control system for position and speed control of a field-oriented induction motor.

REVIEW OF VECTOR CONTROL SYSTEMS                                              197

## 6.7  Examples and Simulations

The examples and simulations below illustrate some of the concepts presented in this chapter. With regard to systems which require precise tuning for optimal operation, it must be mentioned that no significant effort of such tuning was made, the emphasis being on demonstration of principles rather than the quality of operation of these systems.

### Example 6.1.

The example motor operates in steady-state, and under rated conditions, in a vector control system with indirect rotor flux orientation. The inductance parameters of the motor have been overestimated by 25%, and the rotor resistance has been underestimated by 10%.

(a) Determine the per-cent errors of the $i_{DS}^{e*}$, $i_{QS}^{e*}$, and $\omega_r^*$ control commands.

(b) Repeat the calculations if the motor operates in a system with indirect stator flux orientation.

(c) Repeat the calculations if the motor operates in a system with indirect airgap flux orientation.

### Solution

The incorrect values of parameters of the motor used in the control settings of the considered systems are $L_s = 0.053$ H, $L_m = 0.05125$ H, $L_r = 0.05212$ H, $R_r = 0.1404$ Ω, and, consequently, $\tau_r = 0.371$ sec. Coefficients $k_T$, $\sigma$, and $\sigma_r$ depend on the ratio of two inductance parameters, hence they remain unaffected by the incorrect assumption of motor parameters, i.e., $k_T = 1.966$ (rotor flux orientation) or $k_T = 2$ (stator and airgap flux orientation), $\sigma = 0.04925$, and $\sigma_r = 0.01687$. The reference torque, $T^*$, is 61.2 N m.

From Example 4.1, the reference values of the motor fluxes, taken at the rated values, are $\lambda_r^* = 0.6496$ Wb, $\lambda_m^* = 0.6511$ Wb, and

$$\lambda_s^* = |\lambda_m^s + L_{ls}\mathbf{I}_s^s| = |0.6511 e^{j(377t - 94.2°)} + 0.00139 \times 50.5 e^{j(377t - 25.5°)}| = 0.6798 \text{ Wb}.$$

(a) Correct control commands for the system with rotor flux orientation have already been determined in Example 4.4 as $i_{DS}^{e*} = 15.84$ A, $i_{QS}^{e*} =$

47.92 A, and $\omega_r^* = 11.33$ rad/sec. If incorrect values of the motor parameters are used, then the same, correct value of $i_{QS}^{e*}$ is obtained due to the unaffected constant $k_T$ (see Eq. (4.17)). However, from Eq. (4.16), at $p = 0$ (steady state),

$$i_{DS}^{e*} = \frac{\lambda_r^*}{L_m} = \frac{0.6496}{0.05125} = 12.68 \text{ A}$$

which represents an error of **-20%**. In effect, from Eq. (4.15), the slip speed command is

$$\omega_r^* = \frac{1}{0.371} \times \frac{47.92}{12.68} = 10.18 \text{ rad/sec}$$

and the corresponding error is **-10.2%**.

(b) From Eq. (6.15),

$$i_{QS}^{e*} = \frac{3}{6} \times \frac{61.2}{0.6798} = 45.01 \text{ A}.$$

Then, Eq. (6.13) with the correct values of motor parameters is

$$\omega_r^{*2} - \frac{1 - 0.04925}{0.04925^2 \times 0.0424 \times 0.267} \times \frac{0.6798}{45.01} \omega_r^* + \frac{1}{(0.04925 \times 0.267)^2} = 0$$

i.e.,

$$\omega_r^{*2} - 522.9 \omega_r^* + 5783 = 0.$$

Of the two solutions, 511.6 rad/sec and 11.33 rad/sec, of the equation above, it is clearly the latter one that is correct.

Substituting $p = 0$ in Eq. 6.11 gives

REVIEW OF VECTOR CONTROL SYSTEMS

Substituting the correct values of motor parameters in Eqs. (6.65) and

$$i_{DS}^{e*} = \frac{\lambda_s^*}{L_s} + \sigma\tau_r\omega_r^* i_{QS}^{e*} \quad (6.64)$$

from which

$$i_{DS}^{e*} = \frac{0.6798}{0.0424} + 0.04925 \times 0.267 \times 11.33 \times 45.01$$

$$= 22.74 \text{ A}.$$

Note that the same magnitude of the stator current vector has been obtained for both the systems analysed, as $(i_{DS}^{e*})^2 + (i_{QS}^{e*})^2 = (15.84 \text{ A})^2 + (47.92 \text{ A})^2$ (rotor flux orientation) $= (22.74 \text{ A})^2 + (45.01 \text{ A})^2$ (stator flux orientation) $= (50.5 \text{ A})^2$ (see Example 4.1), and also the same slip speed command. This is easy to explain, since in both cases the motor is to operate under the same operating conditions, i.e., with the same stator currents and slip speed.

In a similar way, the $i_{DS}^{e*}$ and $\omega_r^*$ commands following the incorrect assumption of motor parameters can be calculated. The results are $i_{DS}^{e*} = 21.3$ A and $\omega_r^* = 10.3$ rad/sec. The percent errors of the current and slip speed commands are **-6.2 %** and **-9%**, respectively.

(c) From Eq. (6.23),

$$i_{QS}^{e*} = \frac{61.2}{2 \times 0.6511} = 47 \text{ A}.$$

Steady-state equations of the decoupling system for airgap flux orientation, obtained by substituting $p = 0$ in Eqs. (6.30) and (6.31) are

$$i_{DS}^{e*} = \frac{\lambda_m^*}{L_m} + \sigma_r\tau_r\omega_r^* i_{QS}^{e*} \quad (6.65)$$

$$\omega_r^* = \frac{i_{QS}^{e*}}{\tau_r(\frac{\lambda_m^*}{L_m} - \sigma_r i_{DS}^{e*})}. \quad (6.66)$$

Substituting the correct values of motor parameters in Eqs. (6.65) and (6.66) gives

$$i_{DS}^{e*} = \frac{0.6511}{0.041} + 0.01687 \times 0.267 \times 47 \, \omega_r^*$$

$$= 15.88 + 0.212 \, \omega_r^*$$

$$\omega_r^* = \frac{47}{0.267 \left(\frac{0.6511}{0.041} - 0.01687 \, i_{DS}^{e*}\right)}$$

$$= \frac{47}{4.24 - 0.0045 \, i_{DS}^{e*}}.$$

When solved, equations above yield $i_{DS}^{e*} = 17.86$ A and $\omega_r^* = 11.33$ rad/sec. Again, $(i_{DS}^{e*})^2 + (i_{QS}^{e*})^2 = (17.86 \text{ A})^2 + (47 \text{ A})^2 = (50.5 \text{ A})^2$. The values of $i_{DS}^{e*}$ and $\omega_r^*$ when an incorrect assumption of values of motor parameters has been made are 15.74 A and 10.2 rad/sec, respectively. Thus, the $i_{DS}^{e*}$ command error is **-11.9%** and the $\omega_r^*$ command error is **-10%**.

As seen from the results obtained, the systems with stator and airgap flux orientation turned out to be somewhat more robust than that with the rotor flux orientation, particularly with respect to the command of the flux-producing current, $i_{DS}^{e*}$. It must be pointed out, however, that this increased robustness is not the main reason for using the stator or airgap flux vectors as reference vectors, but rather the higher accuracy of practical estimation of these vectors in comparison with that of flux vector of the unaccesible rotor.

All the three systems compared have responded with considerable error of the slip speed, which explains the need for adaptive parameter estimators and tuning systems.

### Example 6.2.

Use Eqs. (6.8) and (6.26), and the correct current and slip speed commands obtained in Example 6.1, to demonstrate the effect of the decoupling systems on the flux control in vector control systems with the stator and airgap flux orientation. Assume steady-state operation of the motor.

# REVIEW OF VECTOR CONTROL SYSTEMS

## Solution

Substituting $p = 0$, $\sigma = 0.04925$, $\tau_r = 0.267$ sec, $\omega_r = 11.33$ rad/sec, $L_s = 0.0424$ H, $i^e_{DS} = 22.72$ A, and $i^e_{QS} = 45.01$ A in Eq. (6.8) gives

$$\lambda^e_{DS} = \frac{1+j0.04925\times0.267\times11.33}{1+j0.267\times11.33}$$

$$\times\ 0.0424(22.72+j45.01) = 0.6798+j0\ \text{Wb}$$

which equals the assumed stator flux command, $\lambda^*_s$. Analogously, substituting $\sigma_r = 0.01687$, $L_m = 0.041$ H, $i^e_{DS} = 17.86$ A, $i^e_{QS} = 47$ A, and the same as before values of $\tau_r$ and $\omega_r$ in Eq. (6.26) yields

$$\lambda^e_{DM} = \frac{1+j0.01687\times0.267\times11.33}{1+j0.267\times11.33}$$

$$\times\ 0.041(17.86+j47) = 0.6496+j0\ \text{Wb}$$

which agrees with the assumed airgap flux command, $\lambda^*_m$.

### Simulation 6.1. Indirect Stator Flux Orientation System

The discontinuous, stepped torque program of Simulations 4.1 through 4.3 is unsuitable for a vector control system with indirect stator flux orientation being a subject of this simulation, because of the differentiation of the torque-producing current, $i^{e*}_{QS}$, in the decoupling system (see Eq. (6.12) and Figure 6.1). Therefore, the reference torque of the simulated example motor follows a different, trapezoidal trajectory. Specifically, the torque developed in the motor is required to linerly increase from zero to 100 N m within the first 0.5 sec, stay constant for the next 1 sec, and decrease, linearly, back to zero within the last 0.5 sec of the operating cycle.

Condition (6.16) for the motor in question is

$$\frac{|T^*|}{\lambda^{*2}_s} \leq \frac{6}{6\times0.0424}\left(\frac{1}{0.04925}-1\right) = 455\ N\ m/Wb^2.$$

The maximum torque required is 100 N m; hence $\lambda^*_s \geq \sqrt{100/455} = 0.47$ Wb. Consequently, the reference value of stator flux was set to 0.5 Wb.

Accurate knowledge of motor parameters is assumed for the first simulation. The torque and stator flux of the motor are shown in Figure 6.25. The reference torque program is realized perfectly, and the stator flux stays constant at the desired level. However, as seen in Figure 6.26, when the inductance parameters are overestimated and the resistance parameters are underestimated, both by 10%, the improperly tuned control system causes transient oscillations of the developed torque and flux.

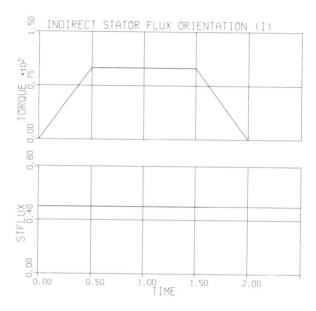

**Figure 6.25.** Torque and stator flux of the motor in the indirect stator flux orientation system with accurate estimation of motor parameters.

To demonstrate the importance of the "torque per flux squared" condition (6.16), the last simulation involves the same system, correctly tuned but with the reference stator flux set to 0.4 Wb, i.e., below the required minimum level. As seen in Figure 6.26, the developed torque initially follows the reference trajectory, but only to the point at which the condition in question is no longer satisfied. Then, the system suddenly becomes unstable, regainig stability only at the end of the operating cycle.

**Simulation 6.2. Vector Control System with Current Source Inverter**

To show the effects of a CSI on operation of a field-oriented induction motor, a system with indirect rotor flux orientation, employing a CSI, is

REVIEW OF VECTOR CONTROL SYSTEMS 203

simulated. The same operating conditions as in Simulation 4.2 are assumed. Torque and speed of the motor are shown in Figure 6.28, stator and rotor current waveforms in Figure 6.29, and rotor flux and torque angle in Figure 6.30.

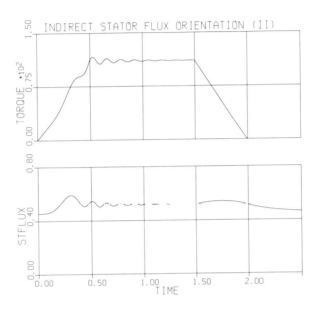

**Figure 6.26.** Torque and stator flux of the motor in the indirect stator flux orientation system with inaccurate estimation of motor parameters.

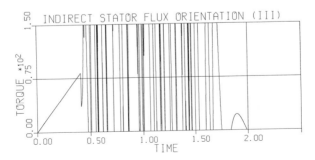

**Figure 6.27.** Torque of the motor in the indirect stator flux orientation system with accurate estimation of motor parameters and insufficient level of stator flux.

Clearly, all the displayed variables of the motor are affected by the discrete, square-wave operation mode of the inverter. However, the average values of the torque, flux, and torque angle follow the required trajectories quite accurately. In effect, the speed trajectory of the drive system does not practically differ from that in Simulation 4.2 in which an ideal source of stator currents was assumed.

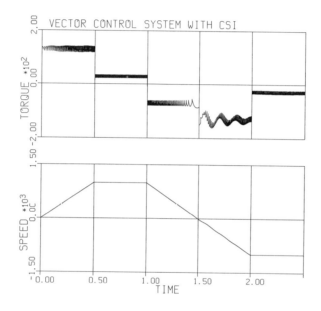

**Figure 6.28.** Torque and speed of the motor in the indirect rotor flux orientation system with CSI.

### Simulation 6.3. Open-Loop Speed Observer

The open-loop observer of the motor speed, described in Section 6.4 and illustrated in Figure 6.12, operates in the vector control system modelled in Simulation 4.3. Both the control system and observer are set to incorrect values of parameters of the example motor. Inductance parameters are overestimated, and resistance parameters are underestimated, all by 10%.

The actual speed and estimated speed of the motor are shown in Figure 6.31. The observer also provides the estimate of the developed torque, compared with the actual torque in Figure 6.32. Figure 6.33 shows the speed and torque errors. Clearly, the accuracy of the observer is far from perfect, which has been the reason for development of closed-loop observers.

# REVIEW OF VECTOR CONTROL SYSTEMS

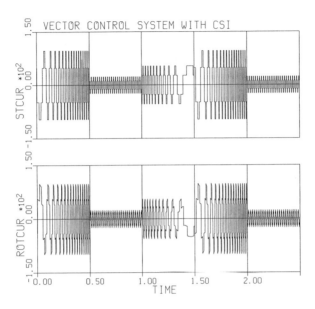

**Figure 6.29.** Stator and rotor currents of the motor in the indirect rotor flux orientation system with CSI.

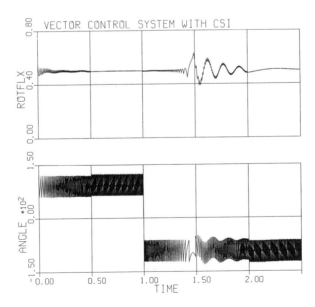

**Figure 6.30.** Rotor flux and torque angle of the motor in the indirect rotor flux orientation system with CSI.

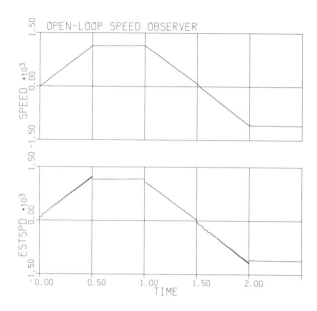

**Figure 6.31.** Actual speed of the motor and speed estimated by the open-loop observer.

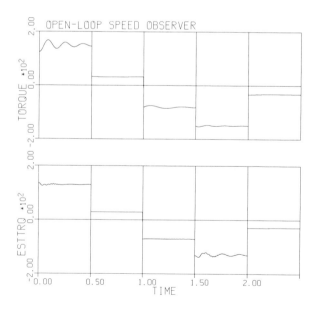

**Figure 6.32.** Actual torque developed by the motor and torque estimated by the open-loop observer.

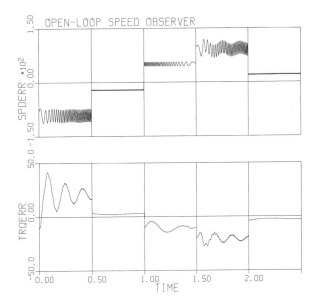

**Figure 6.33.** Speed and torque errors of the open-loop observer.

### Simulation 6.4. Closed-Loop Speed Observer

The same drive system as in Simulation 6.3 is considered, but the open-loop speed observer is replaced with a closed-loop observer of Figure 6.14. The actual and estimated speeds of the motor are shown in Figure 6.34. Certain chattering of the estimated speed, caused by the PI controller, can be discerned. This is particularly visible in Figure 6.35 which shows the speed error. In comparison with the error of the open-loop observer, that of the closed-loop observer is greatly reduced, and the average error is practically zero. Note that the amplitude of the oscillating error is proportional to the absolute speed, which means that the amplitude of a relative error is constant.

### Simulation 6.5. Closed-Loop Rotor Flux Observer

The closed-loop rotor flux observer, described in Section 6.4 and depicted in Figure 6.15, is simulated in conjunction with the drive system from the preceding simulations. The same incorrect values of the motor parameters as in the previous simulations are assumed.

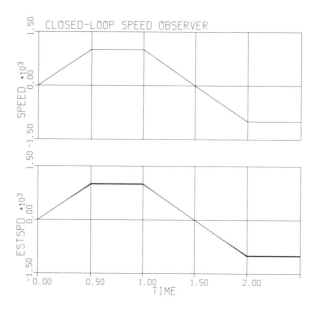

**Figure 6.34.** Actual speed of the motor and speed estimated by the closed-loop observer.

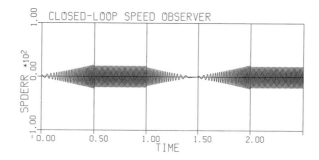

**Figure 6.35.** Speed error of the closed-loop observer.

The actual and estimated magnitudes of the rotor flux vector are shown in Figure 6.36, and the magnitude error in Figure 6.37. The actual and estimated phases of the vector are shown in Figure 6.38. The corresponding phase error is shown in Figure 6.39. It can be seen that the absolute value of flux magnitude error does not exceed 0.065 Wb, i.e., 10% of the average magnitude, and most of the time stays well below this value. The absolute phase error is less than 4°. This level of accuracy fully validates the basic concept of the observer, as the employed value of the rotor time constant, the crucial parameter of the motor, is taken with an error of over 22%.

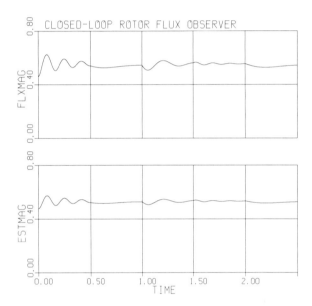

**Figure 6.36.** Actual magnitude of the rotor flux vector and magnitude estimated by the closed-loop observer.

### Simulation 6.6. Adaptive Control of Reference Slip Speed

To illustrate the impact of the adaptive control of reference slip speed, described in Section 6.5, on the operation of a vector control system with indirect rotor flux orientation and incorrect estimation of motor parameters, Simulation 4.3 was repeated, with the scheme shown in Figure 6.20 incorporated into control system of the motor.

The torque and speed of the motor are shown in Figure 6.40, and the rotor flux and torque angle in Figure 6.41. In comparison to those in Figure 4.15, sustained oscillations of the torque, flux, and torque angle can be observed. However, the average values of these quantities are quite close to the reference values. This is illustrated in Figure 6.42 which shows control errors of the torque and flux. For comparison, analogous errors in the system with open-loop control of the reference slip speed (see Simulation 4.3) are shown in Figure 6.43. In effect, the required speed trajectory, 1000 r.p.m. to 0 to -1000 r.p.m., is followed more accurately than in the system in Simulation 4.3. The absolute speed error of the system with adaptive control is less than 40 r.p.m., while that with the open-loop control exceedes 130 r.p.m. because of imperfect realization of the reference torque program.

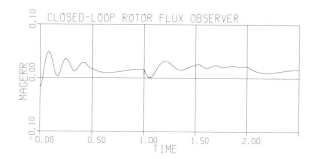

**Figure 6.37.** Magnitude error of the rotor flux vector estimated by the closed-loop observer.

This and the preceding simulations have shown that closed-loop observers and slip speed control systems are superior to their open-loop counterparts. Still, the same simulations have also demonstrated that there is no perfect solution to the problem of parameter uncertainty in vector control systems for induction motors. This is the reason for the intensive pace of research in this area. Although significant achievements have been accomplished in recent years, still the performance of practical systems differs from the ideal operation illustrated in Simulations 4.1 and 4.2. It seems that progress in the area of integrated signal processors will lead to increased implementation of combined schemes of parameter estimation, state observation, and adaptive control in the drive systems based on the FOP.

# REVIEW OF VECTOR CONTROL SYSTEMS

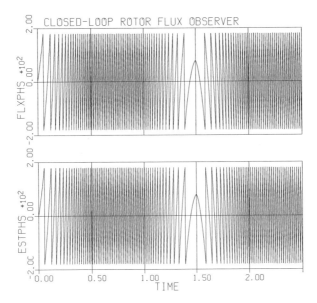

**Figure 6.38.** Actual phase of the rotor flux vector and phase estimated by the closed-loop observer.

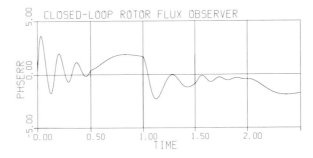

**Figure 6.39.** Phase error of rotor flux vector estimated by the closed-loop observer.

212                                                              CHAPTER 6

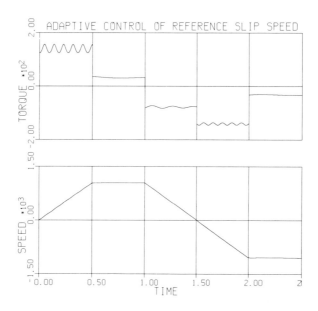

**Figure 6.40.** Torque and speed of the motor in the indirect rotor flux orientation system with adaptive control of reference slip speed.

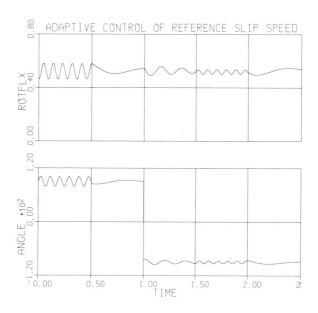

**Figure 6.41.** Rotor flux and torque angle of the motor in the indirect rotor flux orientation system with adaptive control of reference slip speed.

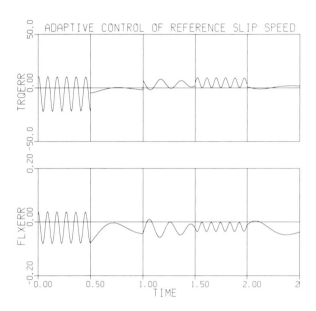

**Figure 6.42.** Control errors of torque and rotor flux in the system with adaptive control of reference slip speed.

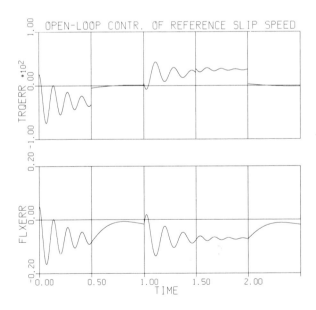

**Figure 6.43.** Control errors of torque and rotor flux in the system with open-loop control of reference slip speed.

## Simulation 6.7. Speed Control Systems

The example motor is required to accelerate within 0.5 sec from a standtill to 800 r.p.m. and maintain this speed independently on the load changes. The load mass moment of inertia is 0.1 kg m$^2$ and the load torque, illustrated in Figure 6.44, changes randomly every 0.1 sec. Its average value increases rapidly after 1.5 sec. The motor operates in the indirect rotor flux orientation system which, for simplicity, is assumed to be ideal, i.e., the developed torque precisely follows the reference torque command.

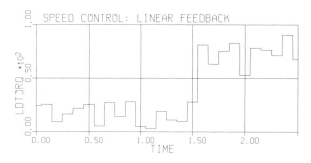

**Figure 6.44.** Variations of the load torque.

A classic, linear-feedback, digital speed control system is simulated first. A linear, PI-type controller is used to generate the reference torque signal in response to the speed error, i.e., the deviation of the actual speed from the desired, semi-trapezoidal trajectory. It should be mentioned that practical controllers are not fully linear, since limits on the torque command must be imposed (here, such limits are not required). The rotor flux is set to 0.5 Wb and the sampling interval of the system is 5 msec.

Speed and speed error of the drive system are shown in Figure 6.45. It can be observed that the variations of the load torque are reflected in the speed error. Because of the idealization of the dynamic model of the system there is no stability-dictated limit on the proportional gain of the controller, so that the speed error could be reduced to any desired extent. However, for the purpose of demonstration, this gain is set at a relatively low level in order to show the effect of the load changes on operation of the system. Similar impact of the load on the developed torque and stator current is illustrated in Figure 6.46.

# REVIEW OF VECTOR CONTROL SYSTEMS

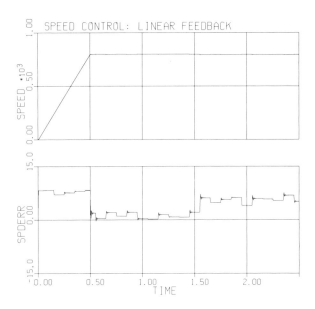

**Figure 6.45.** Speed and speed error of the motor in the linear-feedback speed control system.

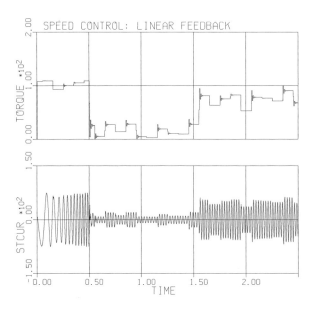

**Figure 6.46.** Torque and stator current of the motor in the linear-feedback speed control system.

The next simulation involves a digital, relay-control, VSC system of Figure 6.22. The same operating conditions of the motor are assumed. The tolerance band for the speed error is set to ±1% of the steady-state speed, i.e., ±8 r.p.m. The switched values of the reference torque, $T^+$ and $T^-$ are 150 N m and zero, respectively, and the sampling interval is 0.5 msec. The results are shown in Figures 6.47 and 6.48. The speed error is now bounded within the tolerance band, i.e., the error is independent on the load, and the average speed precisely follows the desired trajectory. This is, however, accomplished at the expense of chattering, as the motor torque constantly oscillates between the two preassigned values.

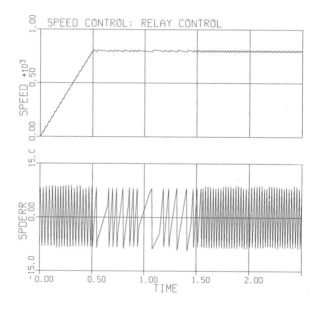

**Figure 6.47.** Speed and speed error of the motor in the relay speed control system.

The last simulation shows how the problem of chattering is alleviated in a digital, combined-control, VSC system of Figure 6.23, with the sampling interval of 5 msec. When the speed error falls within the tolerance band, a linear controller takes over the reference torque control from the relay controller. As a consequence, as seen in Figures 6.49 and 6.50, the relay control is realized, intermittently, only during the acceleration interval. When the steady-state speed has been reached and the speed error stays within the tolerance band, the motor is controlled by the linear controller. In this way, the system combines the advantages of the linear control and relay control, maintaining the speed error bounded within the tolerance band, but with only sporadic chattering.

# REVIEW OF VECTOR CONTROL SYSTEMS

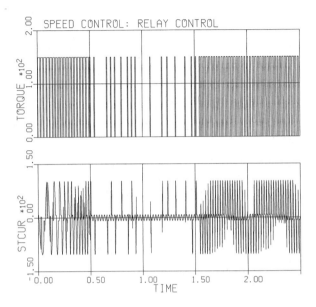

**Figure 6.48.** Torque and stator current of the motor in the relay speed control system.

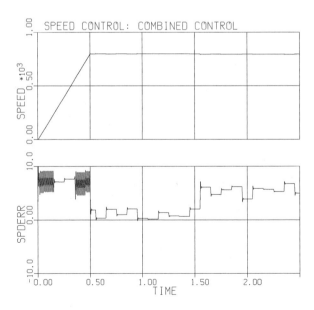

**Figure 6.49.** Speed and speed error of the motor in the combined speed control system.

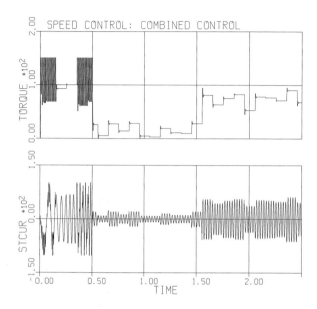

**Figure 6.50.** Torque and stator current of the motor in the combined speed control system.

### Simulation 6.8. Position Control Systems

It can be shown that the least energy is consumed by an electric motor in a positioning system if the speed trajectory is parabolic. Specifically, if the motor is required to change the angular position of its rotor from zero to a certain value, $\Theta_d$, within a specific time interval, $t_d$, then the energy-optimal speed trajectory, $\omega_M(t)$, is

$$\omega_M(t) = 6\frac{\Theta_d}{t_d}(1-\frac{t}{t_d})\frac{t}{t_d} \qquad (6.67)$$

and the corresponding position trajectory, $\Theta_M(t)$, is

$$\Theta_M(t) = 3\Theta_d(1-\frac{2}{3}\frac{t}{t_d})(\frac{t}{t_d})^2. \qquad (6.68)$$

Following the optimal speed trajectory, the motor accelerates to a maximum speed, $\omega_{M,max}$, equal $1.5\Theta_d/t_d$ and, after reaching it at $t = t_d/2$, decelerates to a standstill, arriving at the desired new position at $t = t_d$.

In the subsequent simulations, the load is required to rotate through 5 revolutions in 0.9 sec. A linear-feedback, digital control system of Figure 6.21 is simulated first. The same load is assumed as in the preceding simulation, and the reference position signal, $\Theta_M^*$, is given by Eq. (6.68). PI-type controllers are employed in the inner, speed-control, and outer, position-control, feedback loops. The sampling interval is 0.5 msec.

The position and speed trajectories of the system are shown in Figure 6.51 and the corresponding control errors in Figure 6.52. The developed torque and stator current of the motor are shown in Figure 6.53. It can be seen that the optimal position and speed trajectories are followed quite accurately, although the dynamic position error exceeds, at the middle of the displacement interval, 0.15 of a revolution, i.e., 3% of the total required displacement. Interestingly, the speed trajectory, controlled in an indirect manner, displays the maximum error of 2.5 r.p.m., i.e., 0.5% of the maximum speed only. The impact of the varying load torque on the speed error and torque is easily discernible. As with the linear speed control system in Simulation 6.7, the control errors could be reduced further by increasing the proportional gains of the controllers.

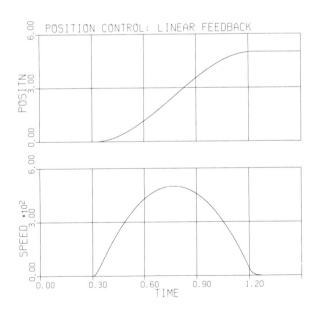

**Figure 6.51.** Position and speed trajectories of the motor in the linear-feedback position control system.

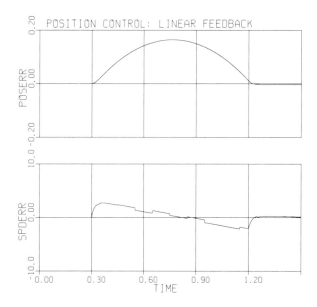

**Figure 6.52.** Position and speed errors of the motor in the linear-feedback position control system.

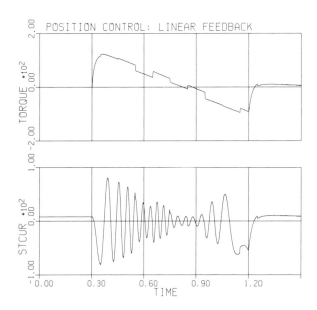

**Figure 6.53.** Torque and stator current of the motor in the linear-feedback position control system.

# REVIEW OF VECTOR CONTROL SYSTEMS

The same operating conditions were assumed for the simulation illustrating a digital, switched-gains, VSC system of Figure 6.24. To reduce the chattering, a tolerance band of ±0.1 of a revolution was employed with respect to the switching function $s$.

The position and speed trajectories are shown in Figure 6.54 and the position and speed errors in Figure 6.55. The effect of the switching mode of operation of the system on the speed trajectory can be seen. This switching mode is also easily observable in Figure 6.56 which depicts the developed torque and stator current waveform. Figure 6.57 shows waveforms of the switched gains, $G_1$ and $G_2$, in the position and speed control loops (see Figure 6.24).

It must be stressed again, that the results obtained should not be interpreted in terms of comparative analysis of quality of the means of control employed. In both cases considered, the control errors can be reduced to any desired level. In general, the switched-gains control system can be expected to be less sensitive to variations of the system parameters and external disturbances than the linear-feedback system. However, the switching mode of operation of the system is associated with undesirable oscillations of the motor speed and may lead to accelerated wear in the mechanical part of the system.

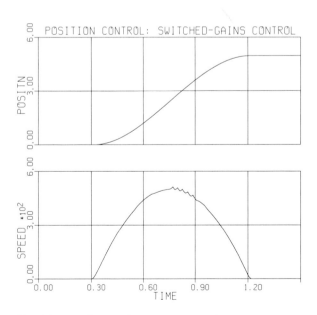

**Figure 6.54.** Position and speed trajectories of the motor in the switched-gains position control system.

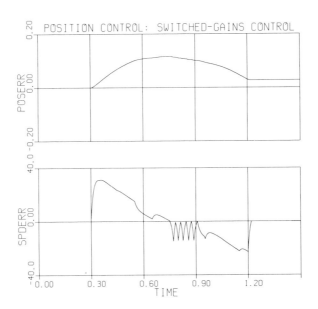

**Figure 6.55.** Position and speed errors of the motor in the switched-gains position control system.

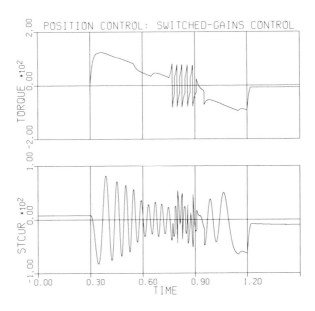

**Figure 6.56.** Torque and stator current of the motor in the switched-gains position control system.

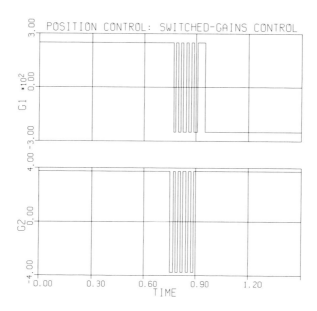

**Figure 6.57.** Waveforms of the switched gains in the position control system.

# Bibliography

The literature on vector control systems with induction motors is very extensive and scores of new papers appear every year. The best English-language sources of current information on the ongoing research and technological advances in the area of vector control of induction motors are:

## Journals

- Electric Machines and Power Systems
- IEEE Transactions on Industry Applications
- IEEE Transactions on Industrial Electronics
- IEEE Transactions on Power Electronics
- Proceedings of the IEE, part B

## Conference Proceedings

- Annual IEEE Industry Applications Society Meeting (IEEE-IAS)
- Annual IEEE Industrial Electronics Society Conference (IECON)
- European Conference on Power Electronics and Applications (EPE)
- IEEE Annual Power Electronics Conference and Exposition (APEC)
- IEEE Power Electronics Specialist Conference (PESC)
- International Conference on Electrical Machines (ICEM)

The nomenclature employed in this book is used by most authors, with possible minor variations in the symbolics. There exists, however, an important difference in the mathematical model of electric machines that the reader will frequently encounter. In many publications, the space vectors of a motor are scaled down by a factor of $1.5\sqrt{2}$ (see Section 1.5 and Eq. (1.60)), so that the magnitude of a given vector equals that of the corresponding phasor. Consequently, the $P/3$ coefficient in torque equations involving vector quantities, such as Eqs. (1.50), (1.90), (3.3), (6.2), and (6.20), is in those publications replaced with $1.5P$ in order to compensate for the downscaling of the involved vectors of current and flux.

The list of selected publications provided below is intended to help the less experienced readers in further, more advanced studies of the subject. With few exceptions, only the books and papers published since 1980 are listed, with the emphasis on the most recent publications. Literature positions which combine several topics appear in the bibliography more than once, under each of the corresponding subheadings.

## Dynamics of Induction Motors and Field Orientation Principle

1. Blashke, F., "The principle of field-orientation as applied to the new 'transvektor' closed-loop control system for rotating-field machines," *Siemens Review*, 34, 5, 1972, pp. 217-220.

2. Boldea, I., and Nasar, S.A., "Electric Machine Dynamics," *Macmillan*, 1986.

3. Boldea, I., and Nasar, S.A., "Vector Control of AC Drives," *CRC Press*, 1992.

4. Bose, B.K., "Power Electronics and AC Drives," *Prentice-Hall*, 1986.

5. Ertem, S., and Baghzouz, Y., "A fast recursive solution for induction motor transients," *IEEE Transactions on Industry Applications*, 24, 5, 1988, pp. 758-764.

6. Gehlot, N.S., and Alsina, P.J., "A discrete model of induction motors for real-time control applications," *IEEE Transactions on Industrial Electronics*, 40, 3, 1993, pp. 317-325.

7. Ghani, S.N., "Digital computer simulation of three-phase induction machine dynamics - A generalized approach," *IEEE Transactions on Industry Applications*, 24, 1, 1988, pp. 106-114.

8. Hasse, K., "About the dynamics of adjustable-speed drives with converter-fed squirrel-cage induction motors" (in German), Dissertation, *Darmstadt Technische Hochschule*, 1969.

9. Hsu, J.S., and Amin, A.M., "Torque calculations of current-source induction machines using the 1-2-0 coordinate system," *IEEE Transactions on Industrial Electronics*, 37, 1, 1990, pp. 34-40.

10. Kawamura, A., and Hoft, R.G., "An analysis of induction motor field oriented or vector control," *Proc. PESC'83*, pp. 91-101.

11. Kovacs, K.P., and Racz, I., "Transient Regimes of AC Machines" (in German), *Hungarian Academy of Sciences*, 1959.

12. Krause, P.C., "Analysis of Electric Machinery," *McGraw-Hill*, 1986.

13. Kron, G., "Equivalent Circuits of Electric Machinery," *John Wiley & Sons*, 1951.

14. Leonhard, W., "Control of Electric Drives," *Springer-Verlag*, 1985.

15. Moreira, J.C., and Lipo, T.A., "Modeling of saturated AC machines including air gap flux harmonic components," *IEEE Transactions on Industry Applications*, **28**, 2, 1992, pp. 343-349.

16. Murata, T., Tsuchiya,T., and Takeda, I., "A new approach for constructing a vector oriented control theory by state space method," *Proc. IECON'87*, pp. 272-279.

17. Murphy, J.M., and Turnbull, F.G., "Power Electronic Control of AC Motors," *Pergamon Press*, 1988.

18. Naunin, D., "The calculation of the dynamic behavior of electric machines by space-phasors," *Electric Machines and Electromechanics*, **4**, 1979, pp. 33-45.

19. Novotny, D.W., and Lorenz, R.D. (Coeditors), "Introduction to Field Orientation and High Performance AC Drives," Tutorial Course, *IEEE Industry Applications Society*, 2nd Ed., 1986.

20. Park, R.H., "Two-reaction theory of synchronous machines - Generalized method of analysis - Part I," *AIEE Transactions*, **48**, 1929, pp. 716-727.

21. Sen, P.C., "Electric motor drives and control - Past, present, and future," *IEEE Transactions on Industrial Electronics*, **37**, 6, 1990, pp. 562-575.

22. Slemon, G.R., "Modelling of induction machines for electric drives," *IEEE Transactions on Industry Applications*, **25**, 6, 1989, pp. 1126-1131.

23. Stanley, H.C., "An analysis of the induction motor," *AIEE Transactions*, **57** (Supplement), 1938, pp. 751-755.

24. Sullivan, C.R., and Sanders, S.R., "Models for induction machines with magnetic saturation in the main flux path," *Conf. Rec. IEEE-IAS'92*, pp. 123-131.

25. Vas, P., Brown, J.E., and Hallenius, K.E., "Small-signal analysis of smooth-airgap machines using space vector theory," *Proc. Electric Energy Conference*, 1987, pp. 210-214.

26. Vas, P., "Vector Control of AC Machines," *Oxford Science Publications*, 1990.

27. Yamamura, S., "AC Motors for High-Performance Applications," *Marcel Dekker*, 1986.

28. Yamamura, S., "Spiral vector theory of AC motor analysis and control," *Conf. Rec. IEEE-IAS'91*, pp. 79-86.

## Scalar Control

29. Abbas, M., and Novotny, D.W, "The stator voltage-controlled current source inverter induction motor drive," *IEEE Transactions on Industry Applications*, **IA-18**, 3, 1982, pp. 219-229.

30. Baader, U., Depenbrock, M., and Gierse, G., "Direct self-control (DSC) of inverter-fed induction machine, basis for speed control without speed-measurement," *IEEE Transactions on Industry Applications*, **28**, 3, 1992, pp. 581-588.

31. Barton, T.H., "Variable frequency variable speed AC drives," *Electric Machines and Power Systems*, **12**, 1987, pp. 143-163.

32. Belmans, R., Geysen, W., and Busschots, F., "Application of field oriented control in crane drives," Conf. Rec. *IEEE-IAS'91*, pp. 347-353.

33. Bose, B.K. (Editor), "Adjustable Speed AC Drive Systems," *IEEE Press*, 1981.

34. Bose, B.K., "Adjustable-speed AC drives - A technology status review," *Proceedings IEEE*, **70**, 1982, pp. 116-135.

35. Bose, B.K., "Scalar decoupled control of induction motor," *IEEE Transactions on Industry Applications*, **IA-20**, 1, 1984, pp. 216-225.

36. Bose, B.K., "Power Electronics and AC Drives," *Prentice-Hall*, 1986.

37. Bowes, S.R., "Steady-state performance of PWM inverter drives," *Proceedings IEE*, B, **130**, 1984, pp. 229-244.

38. Connors, D.P., and Jarc, D.A., "Application considerations for AC drives," *IEEE Transactions on Industry Applications*, **IA-19**, 3, 1983, pp. 455-460.

39. Chauprade, R., and Abbondanti, A., "Variable speed drives: modern concepts and approaches," *Proc. IEEE-IAS Intl. Semiconductor Power*

*Converter Conf.*, 1982, pp. 20-37.

40. Domijan, A., and Embriz-Santander, E., "Harmonic mitigation techniques for the improvement of power quality of adjustable speed drives," *Proc. APEC'90*, pp. 96-108.

41. Espelage, P.M., Nowak, J.M., and Walker, L.H., "Symmetrical GTO current source inverter for wide speed range control of 2300 to 4160 volt, 350 to 7000 hp, induction motors," *Conf. Rec. IEEE-IAS'88*, pp. 302-307.

42. Famouri, P., and Cathey, J.J., "Loss minimization control of an induction motor drive," *IEEE Transactions on Industry Applications*, **27**, 1, 1991, pp. 32-37.

43. Finney, D., "Variable-Frequency AC Motor Drive Systems," *Peter Peregrinus*, 1988.

44. Gastli, A., and Matsui, N., "Stator flux-controlled $V/f$ PWM inverter with identification of IM parameters," *IEEE Transactions on Industrial Electronics*, **39**, 4, 1992, pp. 334-340.

45. Habetler, T.G., and Divan, D.M., "Control strategies for direct torque control using discrete pulse modulation, " *IEEE Transactions on Industry Applications*, **27**, 5, 1991, pp. 893-901.

46. Habetler, T.G., Profumo, F., Pastorelli, M., and Tolbert, L.M., "Direct torque control of induction machines using space vector modulation," *IEEE Transactions on Industry Applications*, **28**, 5, 1992, pp. 1045-1053.

47. Habetler, T.G., Profumo, F., and Pastorelli, M., "Direct torque control of induction machines over a wide speed range," *Conf. Rec. IEEE-IAS'92*, pp. 600-606.

48. Hess, H.L., and Divan, D.M., "Extending the low frequency operation of load commutated inverters with torque control techniques," *Conf. Rec. IEEE-IAS'92*, pp. 607-614.

49. Hickok, H.N., "Adjustable speed - A tool for saving energy losses in pumps, fans, blowers, and compressors," *IEEE Transactions on Industry Applications*, **IA-21**, 1, 1985, pp. 124-136.

50. Hofmann, W., and Krause, M., "Fuzzy control of AC-drives fed by PWM-inverters," Proc. *IECON'92*, pp. 82-87.

51. Huget, E.W., "Squirrel-cage induction motors - Performance versus efficiency," *IEEE Transactions on Industry Applications*, **IA-19**, 5, 1983, pp. 818-823.

52. Inaba, H., Hirasawa, K., Ando, T., Hombu, M., and Nakazato, M., "Development of a high-speed elevator controlled by current source inverter system with sinusoidal input and output," *IEEE Transactions on Industry Applications*, **28**, 4, 1992, pp. 893-899.

53. Ishihara, K., Katayama, T., Watanabe, T., Seto, M., and Matsuyama, I., "AC drive system for tension reel control," *IEEE Transactions on Industry Applications*, **IA-21**, 1, 1985, pp. 147-153.

54. Jarc, D.A., and Connors, D.P., "Variable frequency drives and power factor," *IEEE Transactions on Industry Applications*, **IA-21**, 3, 1985, pp. 771-777.

55. Jarc, D.A., and Novotny, D.W., "A graphical approach to AC drive classification," *IEEE Transactions on Industry Applications*, **IA-23**, 6, 1987, pp. 1029-1035.

56. Jian, T.W., Novotny, D.W., and Schmitz, N.L., "Characteristic induction motor slip values for variable voltage part load performance optimization," *IEEE Transactions on Power Apparatus and Systems*, **PAS-102**, 1, 1983, pp. 38-46.

57. Kazmierkowski, M.P., and Kopcke, H.J., "Comparison of dynamic behaviour of frequency converter fed induction machine drives," *Proc. IFAC Symp. on Control in Power Electronics and Electrical Drives*, 1983, pp. 313-320.

58. Kim, H.G., Sul, S.K., and Park, M.H., "Optimal efficiency drive of a current source inverter fed induction motor by flux control," *IEEE Transactions on Industry Applications*, **IA-20**, 6, 1984, pp. 1453-1459.

59. Kirschen, D.S., Novotny, D.W., and Suwanwisoot, W., "Minimizing induction motor losses by excitation control in variable frequency drives," *IEEE Transactions on Industry Applications*, **IA-20**, 5, 1984, pp. 1244-1250.

60. Kirschen, D.S., Novotny, D.W., and Lipo, T.A., "On line efficiency optimization of a variable fequency induction motor drive," *IEEE Transactions on Industry Applications*, **IA-21**, 4, 1985, pp. 610-615.

61. Koga, K., Ueda, R., and Sonada, T., "Constitution of $V/f$ control for

reducing the steady-state speed error to zero in induction motor drive," *IEEE Transactions on Industry Applications*, **28**, 2, 1992, pp. 463-471.

62. Krishnan, R., Stefanovic, V.R., and Lindsay, J.F., "Control characteristics of inverter-fed induction motor," *IEEE Transactions on Industry Applications*, **IA-19**, 1, 1983, pp. 94-104.

63. Kusko, A., and Galler, D., "Control means for minimization of losses in AC and DC motor drives," *IEEE Transactions on Industry Applications*, **IA-19**, 4, 1983, pp. 561-570.

64. Leffler, F., and King, R.H., "Hoist performance electrical parameters," *Conf. Rec. IEEE-IAS'85*, pp. 192-194.

65. Leonhard, W., "Control of Electric Drives," *Springer-Verlag*, 1985.

66. Leonhard, W., "Adjustable-speed AC drives," *Proceedings IEEE*, **76**, 4, 1988, pp. 455-471.

67. Lockwood, M., "Simulation of unstable oscillations in PWM variable-speed drives," *IEEE Transactions on Industry Applications*, **24**, 1, 1988, pp. 137-141.

68. McConnell, J.E., "The economics for selection of drives - AC motors or steam turbine mechanical drives," *IEEE Transactions on Industry Applications*, **IA-21**, 2, pp. 375-381.

69. Miki, H., Yoshikawa, H., Iwahori, M., Yoshinori, N., and Konishi, Y., "New AC traction drive system with transistor VVVF inverter," *Conf. Rec. IEEE-IAS'91*, pp. 291-297.

70. Mir, S.A., Zinger, D.S., and Elbuluk, M.E., "Fuzzy controller for inverter fed induction machines," *Conf Rec. IEEE-IAS'92*, pp. 464-471.

71. Moreira, J.C., Lipo, T.A., and Blasko, V., "Simple efficiency maximizer for an adjustable frequency induction motor drive," *IEEE Transactions on Industry Applications*, **27**, 5, 1991, pp. 940-946.

72. Murphy, J.M., and Honsinger, V.B., "Efficiency optimization of inverter-fed induction motor drives," *Conf. Rec. IEEE-IAS'82*, pp. 544-552.

73. Murphy, J.M., and Turnbull, F.G., "Power Electronic Control of AC Motors," *Pergamon Press*, 1988.

74. Park, M.H., and Sul, S.K., "Microprocessor-based optimal-efficiency drive of an induction motor," *IEEE Transactions on Industrial Electronics*, **IE-31**, 1, 1984, pp. 69-73.

75. Pottebaum, J.R., "Optimal characteristics of a viariable-frequency centrifugal pump motor drive," *IEEE Transactions on Industry Applications*, **IA-20**, 1, 1984, pp. 23-31.

76. Rao, U.M., Varma, R., Perlekar, S.D., and Acharya, G.N., "Microprocessor based feedback control using reactive power measurement for squirrel cage induction motor," *Conf. Rec. IEEE-IAS'89*, pp. 668-673.

77. Rozner, J.D., Myers, P.W., and Robb, D.J., "The application of adjustable frequency controllers for forced draft fans for improved reliability and energy savings," *IEEE Transactions on Industry Applications*, **IA-21**, 6, pp. 1482-1490.

78. Schauder, C.D., Choo, F.H., and Roberts, M.T., "High performance torque-controlled induction motor drives," *IEEE Transactions on Industry Applications*, **IA-19**, 3, 1983, pp. 74-83.

79. Takahashi, I., and Ohmori, Y., "High performance direct torque control of an induction motor," *IEEE Transactions on Industry Applications*, **25**, 2, 1989, pp. 257-264.

80. Trzynadlowski, A.M., "Computer aided preliminary design of electric drives in the key-parameters space," *Electric Machines and Power Systems*, **12**, 1987, pp. 445-457.

81. Tsuji, T., Iura, H., and Hirata, A., "Vector approximation with DC link current control and identification in AC drive," *Conf. Rec. IEEE-IAS'92*, pp. 563-569.

82. Ueda, R., Sonoda, T., Koga, K., and Ichikawa, M., "Stability analysis in induction motor driven by $V/f$ controlled general-purpose inverter," *IEEE Transactions on Industry Applications*, **28**, 2, 1992, pp. 472-481.

83. Walker, L.H., and Espelage, P.M., "A high-performance controlled-current inverter drive," *IEEE Transactions on Industry Applications*, IA-16, 2, 1980, pp. 193-202.

84. Wu, B., Slemon, G.R., and Dewan, S.B., "PWM-CSI induction motor drive with phase angle control," *IEEE Transactions on Industry*

*Applications,* **27**, 5, 1991, pp. 970-976.

85. Xue, Y., Xu, X., Habetler, T.G., and Divan, D.M., "A stator flux-oriented voltage source variable-speed drive based on DC link measurement," *IEEE Transactions on Industry Applications,* **27**, 5, 1991, pp. 962-969.

86. Zhang, J., and Barton, T.H., "Microprocessor-based primary current control for a cage induction motor drive," *IEEE Transactions on Power Electronics,* **4**, 1, pp. 73-82.

## Inverters

87. Bose, B.K. (Editor), "Adjustable Speed AC Drive Systems," *IEEE Press,* 1981.

88. Bose, B.K., "Power Electronics and AC Drives," *Prentice-Hall,* 1986.

89. Bose, B.K., "Power electronics - An emerging technology," *IEEE Transactions on Industrial Electronics,* **36**, 3, 1989, pp. 403-412.

90. Bose, B.K., "An adaptive hysteresis-band current control technique of a voltage-fed PWM inverter for machine drive system," *IEEE Transactions on Industrial Electronics,* **37**, 5, 1990, pp. 402-408.

91. Bose, B.K., "Power electronics - A technology review," *Proceedings IEEE,* **80**, 8, 1992, pp. 1303-1334.

92. Bose, B.K., "Recent advances in power electronics," *IEEE Transactions on Power Electronics,* **7**, 1, 1992, pp. 2-16.

93. Bose, B.K. (Editor), "Modern Power Electronics: Evolution, Technology, and Applications," IEEE Press, 1992.

94. Bowes, S.R., and Clark, P.R., "Transputer-based optimal PWM control of inverter drives," *IEEE Transactions on Industry Applications,* **28**, 1, 1992, pp. 81-88.

95. Boys, J.T., and Handley, P.G., "Spread spectrum switching: low noise modulation technique for PWM inverter drives," *Proceedings IEE,* B, **139**, 3, 1992, pp. 252-260.

96. Brichant, F., "Force-Commutated Inverters," *Macmillan,* 1982.

97. Brod, D.M., and Novotny, D.W., "Current control of VSI-PWM inverters, "*IEEE Transactions on Industry Applications*, IA-21, 3, 1985, pp. 562-570.

98. Dewan, S.B., Slemon, G.R., and Straughen, A., "Power Semiconductor Drives," *John Wiley & Sons*, 1984.

99. Enjeti, P.N., Ziogas, P.D., and Lindsay, J.F., "Programmed PWM techniques to eliminate harmonics: a critical evaluation," *IEEE Transactions on Industry Applications*, **26**, 2, 1990, pp. 302-316.

100. Enjeti, P.N., Ziogas, P.D., and Lindsay, J.F., "A current source PWM inverter with instantaneous current control capability," *IEEE Transactions on Industry Applications*, **27**, 3, 1991, pp. 582-588.

101. Enjeti, P.N., Ziogas, P.D., Lindsay, J.F., and Rashid, M.H., "A new current control scheme for AC motor drives," *IEEE Transactions on Industry Applications*, **28**, 4, 1992, pp. 842-849.

102. Erdman, W.L., Hudson, R., Yang, J., and Hoft, R.G., "A 7.5-kW ultrasonic inverter motor drive employing MOS-controlled thyristors," *IEEE Transactions on Industry Applications*, **26**, 4, 1990, pp. 756-768.

103. Espinoza, J., Joos, G., and Ziogas, P., "Voltage controlled current source inverters," Proc. IECON'92, pp. 512-517.

104. Espelage, P.M., Nowak, J.M., and Walker, L.H., "Symmetrical GTO current source inverter for wide speed range control of 2300 to 4160 volt, 350 to 7000 hp, induction motors," *Conf. Rec. IEEE-IAS'88*, pp. 302-307.

105. Gosbell, V.J., and Dalton, P.M., "Current control of induction motors at low speeds," *IEEE Transactions on Industry Applications*, **28**, 2, 1992, pp. 482-489.

106. Habetler, T.G., and Divan, D.M., "Acoustic noise reduction in sinusoidal PWM drives using a randomly modulated carrier," *IEEE Transactions on Power Electronics*, **6**, 3, 1991, pp. 356-363.

107. Hess, H.L., and Divan, D.M., "Extending the low frequency operation of load commutated inverters with torque control techniques," *Conf. Rec. IEEE-IAS'92*, pp. 607-614.

108. Heumann, K., "Basic Principles of Power Electronics," *Springer-*

*Verlag*, 1986.

109. Hoft, R.G., "Semiconductor Power Electronics," *Van Nostrand Reinhold*, 1986.

110. Holtz, J., "Pulsewidth modulation - A survey," *IEEE Transactions on Industrial Electronics*, **39**, 5, 1992, pp. 410-420.

111. Holtz, J. and Bube, E., "Field-oriented asynchronous pulse-width modulation for high-performance AC machine drives operating at low switching frequency," *IEEE Transactions on Industry Applications*, **27**, 3, 1991, pp. 574-581.

112. Kassakian, J.G., Schlecht, M.F., and Verghese, G.C., "Principles of Power Electronics," *Addison-Wesley*, 1991.

113. Kazmierkowski, M.P., Dzieniakowski, M.A., and Sulkowski, W., "Novel space vector based current controllers for PWM inverters," *IEEE Transactions on Power Electronics*, **6**, 1, 1991, pp. 158-166.

114. Kerkman, R.J., and Rowan, T.M., "Voltage-contrlled, current-regulated PWM Inverters, *IEEE Transactions on Industry Applications*, **26**, 2, 1990, pp. 244-251.

115. Kerkman, R.J., Seibel, B.J., Brod, D.M., Rowan, T.M., and Leggate, D., "A simplified inverter model for on-line control and simulation," *IEEE Transactions on Industry Applications*, **27**, 3, 1991, pp. 567-573.

116. Legowski, S., and Trzynadlowski, A.M., "Minimum-loss vector PWM strategy for three-phase inverters," *Proc. APEC'93*, pp. 785-792.

117. Malesani, L., Tenti, P., Gaio, E., and Piovan, R., "Improved current control technique of VSI PWM inverters with constant modulation frequency and extended voltage range," *IEEE Transactions on Industry Applications*, **27**, 2, 1991, pp. 365-369.

118. Matsue, K., and Kubota, H., "Analysis of PWM current source GTO inverter with PWM-controlled thyristor rectifier for induction motor drive," *IEEE Transactions on Industry Applications*, **26**, 2, 1990, pp. 274-282.

119. Mohan, N., Undeland, T.M., and Robbins, W.P., "Power Electronics: Converters, Applications, and Design," *John Wiley & Sons*, 1989.

120. Nabae, A., Ogasawara, S., and Akagi, H., "A novel scheme for

current-controlled PWM inverters," *IEEE Transactions on Industry Applications*, **IA-22**, 4, 1986, pp. 697-701.

121. Nonaka, S., and Neba, Y., "A PWM GTO current source converter-inverter system with sinusoidal inputs and outputs," *Conf. Rec. IEEE-IAS'87*, pp. 247-252.

122. Rashid, M.H., "Power Electronics: Circuits, Devices, and Applications," *Prentice Hall*, 1993.

123. Rowan, T.R., and Kerkman, R.J., "A new synchronous current regulator and an analysis of current-regulated PWM inverters," *IEEE Transactions on Industry Applications*, **IA-22**, 4, pp. 678-690.

124. Shepherd, W., and Hulley, L.N., "Power Electronics and Motor Control," *Cambridge University Press*, 1987.

125. Tripathi, A., and Sen, P.C., "Comparative analysis of fixed and sinusoidal band hysteresis current controller for voltage source inverters," *IEEE Transactions on Industrial Electronics*, **39**, 1, pp. 63-73.

126. Trzynadlowski, A.M., Legowski, S., and Kirlin, R.L., "Random pulse width modulation technique for voltage-controlled power inverters," *Conf. Rec. IEEE-IAS'87*, pp. 863-868.

127. Van der Broeck, H.W., Skudelny, H.C., and Stanke, G.V., "Analysis and realization of a pulsewidth modulator based on voltage space vectors," *IEEE Transactions on Industry Applications*, **IA-24**, 1, 1988, pp. 142-150.

128. Wu, B., Dewan, S.B., and Slemon, G.R., "PWM-CSI induction motor drive with phase angle control," *Conf. Rec. IEEE-IAS'89*, pp. 674-679.

129. Wu, B., Dewan, S.B., and Slemon, G.R., "PWM-CSI inverter for induction motor drives," *IEEE Transactions on Industry Applications*, **28**, 1, 1992, pp. 64-71.

## Vector Control

130. Bassi, E., Benzi, F.P., Bolognani, S., and Buja, G.S., "A field orientation scheme for current-fed induction motor drives based on the torque angle closed-loop control," *IEEE Transactions on Industry Applications*, **28**, 5, 1992, pp. 1038-1044.

131. Bausch, H., and Hontheim, H., "Transient performance of induction machines with field-oriented control," *Proc. ICEM'85*, pp. 199-203.

132. Bausch, H., Hontheim, H., and Kolletschke, D., "The influence of decoupling methods on the dynamic behaviour of a field-oriented controlled induction machine," *Proc. ICEM'86*, pp. 648-651.

133. Bose, B.K., "Power Electronics and AC Drives," *Prentice-Hall*, 1986.

134. Bellini, A., Figalli, G., and Ulivi, G.,"A microprocessor based direct field oriented control of induction motors," *Proc. ICEM'86*, pp. 652-665.

135. Ben-Brahim, L., and Kawamura, A., "A fully digitized field-oriented controlled induction motor drive using only current sensors," *IEEE Transactions on Industrial Electronics*, **39**, 3, 1992, pp. 241-249.

136. Bodson, M., and Chiasson, J., "A systematic approach to selecting optimal flux references in induction motors," *Conf. Rec. IEEE-IAS'92*, pp. 531-537.

137. Boldea, I., and Nasar, S.A., "Vector Control of AC Drives," *CRC Press*, 1992.

138. Bolognani, S., and Buja, G.S., "Parameter variation and computation error effects in indirect field-oriented induction motor drives," *Proc. ICEM'88*, pp. 545-549.

139. Chen, S., and Yeh, S.N., "Efficiency control of field oriented operation based on the open-loop VVVF drivers," *Conf. Rec. IEEE-IAS'91*, pp. 58-64.

140. Dalton, P.M., and Gosbell, V.J., "30 kW asynchronous drive with improved response to rotor resistance variation," *Conf. Rec. IEEE-AS'91*, pp. 469-474.

141. Deng,D., and Lipo, T.A., "A modified control method for fast response current source inverter drives," *IEEE Transactions on IndustryApplications*, **IA-22**, 4, 1986, pp. 653-665.

142. De Doncker, R., Vandenput, A., and Geysen, W., "A digital field-oriented controller using the double-cage induction motor model," *Proc. PESC'86*, pp. 502-509.

143. De Doncker, R., and Novotny, D.W., "The universal field-oriented controller," *Conf. Rec. IEEE-IAS'88*, pp. 450-456.

144. De Doncker, R., "Field oriented controllers with rotor deep bar compensation circuits," *IEEE Transactions on Industry Applications*, **28**, 5, 1992, pp. 1062-1071.

145. Ferraris, P., Fratta, A., Vagati, A., and Vilatta, F., "About the vector control of induction motors for special applications without speed sensor," *Proc. Intl. Conf. on the Evolution and Modern Aspects of Induction Machines*, 1986, pp. 444-450.

146. Fratta, A., Vagati, A., and Vilatta, F., "Vector control of induction motors without transducers," *Proc. PESC'88*, pp. 839-846.

147. Gabriel, R., Leonhard, W., and Nordby, C., "Field-oriented control of a standard AC motor using microprocessors," *IEEE Transactions on Industry Applications*, **IA-16**, 2, 1980, pp. 186-192.

148. Garcia, G.O., Mendes Luis, J.C., Stephan, R.M., and Watanabe, E.H., "Fast efficiency maximizer for adjustable speed induction motor drive," *Proc. IECON'92*, pp. 37-42.

149. Harashima, F., Kondo, S., Ohnishi, K., Kajita, M., and Susono, M., "Multimicroprocessor-based control system for quick response induction motor drive," *IEEE Transactions on Industry Applications*, **IA-21**, 3, 1985, pp. 602-609.

150. Harashima, F., and Arihara, E., "Microcomputer-controlled induction motor considering the effects of secondary resistance variation," *Conf. Rec. IEEE-IAS'85*, pp. 548-553.

151. Harley, R.G., Hemme, A.W., Levy, D.C., and Webster, M.R., "Real-time issues of transputers in high-performance motion control systems," *IEEE Transactions on Industry Applications*, **29**, 2, 1993, pp. 306-312.

152. Ho, E.Y., and Sen, P.C., "Decoupling control of induction motor drives," *IEEE Transactions on Industrial Electronics*, **35**, 2, 1988, pp. 253-262.

153. Holtz, J., and Stadtfeld, S., "Field-oriented control by forced motor currents in a volatge-fed induction motor drive," *Proc. IFAC Symp. on Control in Power Electronics and Electrical Drives*, 1983, pp. 103-110.

154. Hyun, D.S., Cho, S.B., and Lee, T.K., "Sensorless vector control of induction motor compensating the variation of rotor resistance," *Proc.*

*IECON'92*, pp. 72-76.

155. Irisa, T., Takata, S., Ueda, R., and Sonoda, T., "Effect of machine structure identification of spatial position and magnitude of rotor flux in induction motor vector control," *Proc. IEEE Conf. on Power Electronics and Variable Speed Drives*, 1984, pp. 352-356.

156. Islam, S.M., and Somouah, C.B., "An efficient high performance voltage decoupled induction motor drive with excitation control," *IEEE Transactions on Energy Conversion*, **4**, 1, 1989.

157. Jarc, D.A., and Novotny, D.W., "A graphical approach to AC drive classification," *IEEE Transactions on Industry Applications*, **IA-23**, 6, 1987, pp. 1029-1035.

158. Joetten, R., and Mader, G., "Control methods for good dynamic performance induction motor drives based on current and voltage as measured quantities," *IEEE Transactions on Industry Applications*, **IA-19**, 3, 1983, pp. 356-362.

159. Kao, Y.T., and Liu, C.H., "Analysis and design of microprocessor-based vector-controlled induction motor drives," *IEEE Transactions on Industrial Electronics*, **39**, 1, 1992, pp. 46-54.

160. Kazmierkowski, M.P., and Kopcke, H.J., "Comparison of dynamic behaviour of frequency converter fed induction machine drives," *Proc. IFAC Symp. on Control in Power Electronics and Electrical Drives*, 1983, pp. 313-320.

161. Kazmierkowski, M.P., and Kopcke, H.J., "A simple control system for current source inverter-fed induction motor drives," *IEEE Transactions on Industry Applications*, **IA-21**, 3, 1985, pp. 617-623.

162. Kazmierkowski, M.P., and Sulkowski, W., "A novel vector control scheme for transistor PWM inverter-fed induction motor drive," *IEEE Transactions on Industrial Electronics*, **38**, 1, 1991, pp. 41-47.

163. Kerkman, R.J., Rowan, T.M., and Legatte, D., "Indirect field-oriented control of an induction motor in the field-weakening region," *IEEE Transactions on Industry Applications*, **28**, 4, 1992, pp. 850-857.

164. Khater, F.M., Lorenz, R.D., Novotny, D.W., and Tang, K., "Selection of flux level in field-oriented induction machine controllers with consideration of magnetic saturation effects," *IEEE Transactions on Industry Applications*, **IA-23**, 2, 1987, pp. 276-282.

165. Kohlmeier, H., Niermeyer, O., and Schroder, D.F., "Highly dynamic four-quadrant AC motor drive with improved power factor and on-line optimized pulse pattern with PROMC," *IEEE Transactions on Industry Applications*, **IA-23**, 6, 1987, pp. 1001-1009.

166. Kreindler, L., Moreira, J.C., Testa, A., and Lipo, T.A., "Direct field orientation controller using the stator phase voltage third harmonic," *Conf. Rec. IEEE-IAS'92*, pp. 508-514.

167. Krishnan, R., and Doran, F.C., "Study of parameter sensitivity in high-performance inverter-fed induction motor drive systems," *IEEE Transactions on Industry Applications*, **IA-23**, 4, 1987, pp. 623-635.

168. Kume, T., and Iwakane, T., "High performance vector controlled AC motor drives: applications and new techniques," *Conf. Rec. IEEE-IAS'85*, pp. 690-697.

169. Kume, T., Iwakane T., Sawa, T., Yoshida, T., and Nagai, I., "A wide constant power range vector-controlled AC motor drive using winding changeover technique," *IEEE Transactions on Industry Applications*, **27**, 5, 1991, pp. 934-939.

170. Kume, T., Sawa, T., Yoshida, T., Sawamura, M., and Sakamoto, M., "A high-speed vector-controlled spindle motor drive with closed transition between with encoder control and without encoder control," *IEEE Transactions on Industry Applications*, **28**, 2, 1992, pp. 421-426.

171. Lee, D.C., Sul, S.K., and Park, M.H., "High performance current regulator for a field oriented controlled induction motor," *Conf. Rec. IEEE-IAS'92*, pp. 538-544.

172. Leonhard, W., "Control of Electric Drives," *Springer-Verlag*, 1985.

173. Leonhard,W., "Microcomputer control of high dynamic performance AC drives - A survey," *Automatica*, **22**, 1986, pp. 1-19.

174. Leonhard, W., "Adjustable-speed AC drives," *Proceedings IEEE*, **76**, 4, 1988, pp. 455-471.

175. Leonhard, W., "Field-orientation for controlling AC machines - Principle and application," *Proc. Intl. Conf. on Power Electronics and Variable Speed Drives*, 1988, pp. 277-282.

176. Levi, E., and Vuckovic, V., "Field oriented control of induction

machines in the presence of saturation," *Electric Machines and Power Systems*, **16**, 1989, pp. 133-147.

177. Li, W., "A highly reliable parallel processing controller for vector control of AC induction motor," *Proc. IECON'92*, pp. 43-48.

178. Liaw, C.M., Kung, Y.S., and Wu, C.M., "Design and implementation of a high-performance field oriented induction motor drive," *IEEE Transactions on Industrial Electronics*, **38**, 4, 1991, pp. 275-282.

179. Lipo, T.A., and Chang, K.C., "A new approach to flux and torque sensing in induction machines," *IEEE Transactions on Industry Applications*, **IA-22**, 4, 1986, pp. 731-737.

180. Lipo, T.A., Liu, T.H., and Fu, J.R., "A strategy for improving reliability of field oriented controlled induction motor drives," *Conf. Rec. IEEE-IAS'91*, pp. 449-455.

181. Liu, C.H., Hwu, C.C., and Feng, Y.F., "Modeling and implementation of a microprocessor-based CSI-fed induction motor drive using field-oriented control," *IEEE Transactions on Industry Applications*, **25**, 4, 1989, pp. 588-597

182. Lorenz, R.D., "Tuning of field-oriented induction motor controllers for high-performance applications," *IEEE Transactions on Industry Applications*, **IA-22**, 2, 1986, pp. 293-297.

183. Lorenz, R.D., Lucas, M.O., and Lawson, D.B., "Synthesis of a state-variable motion controller for high-performance field oriented induction machine drives," *Conf. Rec. IEEE-IAS'86*, pp. 80-85.

184. Lorenz, R.D., and Novotny, D.W., "Saturation effects in field-oriented induction machines," *IEEE Transactions on Industry Applications*, **26**, 2, 1990, pp. 283-289.

185. Lorenz,R.D., and Lawson, D.B., "Flux and torque decoupling control for field-weakened operation of field-oriented induction machines," *IEEE Transactions on Industry Applications*, **26**, 2, 1990, pp. 290-295.

186. Lorenz, R.D., and Divan, D.M., "Dynamic analysis and experimental evaluation of delta modulators for field-oriented induction machines," *IEEE Transactions on Industry Applications*, **26**, 2, 1990, pp. 296-301.

187. Lorenz, R.D., and Yang, S.M., "Efficiency-optimized flux trajectories

for closed-cycle operation of field-orientation induction machine drives," *IEEE Transactions on Industry Applications*, **28**, 3, 1992, pp. 574-580.

188. Miki, I., Nakao, O., and Nishiyama, S., "A new simplified current control method for field-oriented induction motor drives," *IEEE Transactions on Industry Applications*, **27**, 6, 1991, pp. 1081-1086.

189. Murata, T., Tsuchiya, T., and Takeda, I., "Vector control for induction machine on the application of optimal control theory," *IEEE Transactions on Industrial Electronics*, **37**, 4, 1990, pp. 283-290.

190. Murphy, J.M., and Turnbull, F.G., "Power Electronic Control of AC Motors," *Pergamon Press*, 1988.

191. Nordin, K.B., Novotny, D.W., and Zinger, D.S., "The influence of motor parameter deviations in feedforward field orientation drive systems," *IEEE Transactions on Industry Applications*, **IA-21**, 4, 1985, pp. 1009-1015.

192. Novotny, D.W., and Lorenz, R.D. (Coeditors), "Introduction to Field Orientation and High Performance AC Drives," Tutorial Course, *IEEE Industry Applications Society*, 2nd Ed., 1986.

193. Ohm, D., and Khersonsky, Y., "Operating characteristics of the indirect field oriented induction motor control," *Proc. APEC'89*, pp. 91-100.

194. Ohnishi, K., Suzuki, H., Miyachi, K., and Terashima, M., "Decoupling control of secondary flux and secondary current in induction motor drive with controlled voltage source and its comparison with volts/hertz control," *IEEE Transactions on Industry Applications*, **IA-21**, 1, 1985, pp. 241-247.

195. Osman, R.H., and Bange, J.B., "A regenerative centrifuge drive using a current-fed inverter with vector control," *IEEE Transactions on Industry Applications*, **27**, 6, 1991, pp. 1076-1080.

196. Profumo, F., Grivia, G., Pastorelli, M., Moreira, J., and De Doncker, R., "Universal field oriented controller based on air gap flux sensing via third harmonic stator voltage," *Conf. Rec. IEEE-IAS'92*, pp. 515-523.

197. Sathikumar, S., and Vithayathil, J., "Digital simulation of field-oriented control of induction motor," *IEEE Transactions on Industrial*

*Electronics*, **IE-31**, 2, 1984, pp. 141-148.

198. Sen, P.C., "Electric motor drives and control - Past, present, and future," *IEEE Transactions on Industrial Electronics*, **37**, 6, 1990, pp. 562-575.

199. Stefanovic,V.R., and Nelms, R.M. (Coeditors), "Microprocessor Control of Motor Drives and Power Converters," Tutorial Course, *IEEE-IAS*, 1992.

200. Steinke, J.K., Dudler, G.J., and Huber, B.P., Field oriented control of a high power GTO-VSI fed AC drive with high dynamic performance using a programmable high speed controller (PHSC)," *Conf. Rec. IEEE-IAS'92*, pp. 393-399.

201. Saito, K., Kamiyama, K., Sukegawa, T., Matsui, T., and Okuyama, T., "A multi-processor based, fully digital AC drive system for rolling mills," *IEEE Transactions on Industry Applications*, **IA-23**, 3, 1987, pp. 538-544.

202. Sul, S.K., and Lipo, T.A., "Field-oriented control of an induction machine in a high frequency link power system," *IEEE Transactions on Power Electronics*, **5**, 4, 1990, pp. 436-445.

203. Tajima, I., and Hori, Y., "Speed sensorless field-orientation control of the induction machine," *IEEE Transactions on Industry Applications*, **29**, 1, 1993, pp. 175-180.

204. Takahashi, I., and Noguchi, T., "A new quick response and high efficiency strategy of an induction motor," *Conf. Rec. IEEE-IAS'85*, pp. 495-502.

205. Takahashi, I., and Ohmori, Y., "High performance direct torque control of an induction machine," *IEEE Transactions on Industry Appplications*, **25**, 2, 1989, pp. 257-264.

206. Tsuji, M., Yamada, E., Izumi, K., and Oyama, J., "Stability analysis of a current-source inverter-fed induction motor under vector control," *Proc. ICEM'84*, pp. 867-874.

207. Vas, P., "Vector Control of AC Machines," *Oxford Science Publications*, 1990.

208. Wallace, I.T., Lorenz, R.D., Novotny, D.W., and Divan, D.M., "Increasing the dynamic torque per ampere capability of induction machines," *Conf. Rec. IEEE-IAS'91*, pp. 14-20.

209. Wallace, I.T., Novotny, D.W., Lorenz, R.D., and Divan, D.M., "Verification of enhanced dynamic torque per ampere capability of saturated induction machines," *Conf. Rec. IEEE-IAS'92*, pp. 40-47.

210. Wu, Z.K., and Strangas, E.G., "Feedforward field orientation control of an induction motor using a PWM voltage source inverter and standardized single-board computers," *IEEE Transactions on Industrial Electronics*, 35, 1, 1988, pp. 75-79.

211. Xu, X., De Doncker, R., and Novotny, D.W., "Stator flux-orientation control of induction machines in the field-weakening region," *Conf. Rec. IEEE-IAS'88*, pp. 437-443.

212. Xu, X., De Doncker, R., and Novotny, D.W., "A stator flux oriented induction machine drive," *Proc. PESC'88*, pp. 870-876.

213. Xu, X., and Novotny, D.W., "Implementation of direct stator flux orientation control on a versatile DSP based system," *IEEE Transactions on Industry Applications*, 27, 4, 1991, pp. 694-700.

214. Xu, X., and Novotny, D.W., "Selection of the flux reference for induction machine drives in the field weakening region," *IEEE Transactions on Industry Applications*, 28, 6, 1992, pp. 1353-1358.

215. Zhang, J., Thiagaranjan, V., Grant, T., and Barton, T.H., "New approach to field orientation control of a CSI induction motor drive," *Proceedings IEE*, B, 135, 1988, pp. 1-7.

216. Zhang, J., and Barton, T.H., "A fast variable structure current controller for an induction machine drive," *IEEE Transactions on Industry Applications*, 26, 3, 1990, pp. 415-419.

217. Zinger, D.S., Profumo, F., Lipo, T.A., and Novotny, D.W., "A direct field oriented controller for induction motor using tapped stator windings," *IEEE Transactions on Power Electronics*, 5, 4, 1990, pp. 446-453.

218. Zinger, D.S., "A flux control for improved dynamic response," *Conf. Rec. IEEE-IAS'91*, pp. 47-51.

## Observers, Parameter Estimation, and Adaptive Schemes

219. Atkinson, D.J., Acarnley, P.P., and Finch, J.W., "Observers for induction motor state and parameter estimation," *IEEE Transactions*

*on Industry Applications*, **27**, 6, 1991, pp. 1119-1127.

220. Bottura, C.P., Silvino, J.L., and de Resende, P., "A flux observer for induction machines based on a time-variant discrete model," *IEEE Transactions on Industry Applications*, **29**, 2, 1993, pp. 349-354.

221. Chan, C.C., Leung, W.S., and Ng, C.W., "Adaptive decoupling control of induction motor drives," *IEEE Transactions on Industrial Electronics*, **37**, 1, 1990, pp. 41-47.

222. Chan, C.C., and Wang, H., "An effective method for rotor resistance identification for high-performance induction motor vector control," *IEEE Transactions on Industrial Electronics*, **37**, 6, 1990, pp. 477-482.

223. Chrzan, P.J., and Kurzynski, P., "A rotor time constant evaluation for vector-controlled induction motor drives," *IEEE Transactions on Industrial Electronics*, **39**, 5, 1992, pp. 463-465.

224. Dalton, P.M., and Gosbell, V.J., "Flux tracking in induction machines by means of volt-ampere quantities," *IEEE Transactions on Industry Applications*, **26**, 1, 1990, 137-142.

225. Dalton, P.M., and Gosbell, V.J., "30 kW asynchronous drive with improved response to rotor resistance variation," *Conf. Rec. IEEE-IAS'91*, pp. 469-474.

226. De Doncker, R., Vandenput, A., and Geysen, W., "Thermal models of inverter fed machines suited for adaptive temperature compensation of field oriented controllers," *Conf. Rec. IEEE-IAS'86*, pp. 132-139.

227. Holtz, J., and Thimm, T., "Identification of machine parameters in a vector-controlled induction motor drive," *IEEE Transactions on Industry Applications*, **27**, 6, 1991, pp. 1111-1118.

228. Hori, Y., and Umeno, T., "Implementation of robust flux observer-based field orientation (FOFO) controller for induction machines," *Conf. Rec. IEEE-IAS'89*, pp. 523-528.

229. Hyun, D.S., Cho, S.B., and Lee, T.K., "Sensorless vector control of induction motor compensating the variation of rotor resistance," *Proc. IECON'92*, pp. 72-76.

230. Jansen, P.L., and Lorenz, R.D., "A physically insightful appraoch to the design and accuracy assessment of flux observers for field oriented induction machine drives," *Conf. Rec. IEEE-IAS'92*, pp. 570-

577.

231. Kazmierkowski, M.P., and Kopcke, H.J., "A simple control system for current source inverter-fed induction motor drives," *IEEE Transactions on Industry Applications*, **IA-21**, 3, 1985, pp. 617-623.

232. Khambadkone, A.M., and Holtz, J., "Vector controlled induction motor drive with a self-commissioning scheme," *IEEE Transactions on Industrial Electronics*, **38**, 5, 1991, pp. 322-327.

233. Kim, Y.R., Sul, S.K., and Park, M.H., "Speed sensorless vector control of an induction motor using an extended Kalman filter," *Conf. Rec. IEEE-IAS'92*, pp. 594-599.

234. Klaes, N.R., "Parameter identification of an induction machine with regard to dependencies on saturation," *Conf. Rec. IEEE-IAS'91*, pp. 21-27.

235. Koyama, M., Yano, M., Kamiyama, I., and Yano, S., "Microprocessor-based vector control system for induction motor drives with rotor time constant identification," *IEEE Transactions on Industry Applications*, **IA-22**, 3, 1986, pp. 453-459.

236. Kubota, H., and Matsuse, K., "Robust field oriented motor drives based on disturbance torque estimation without rotational transducers," *IEEE-IAS'92*, pp. 558-562.

237. Kubota, H., and Matsuse, K., "Compensation for core loss of adaptive flux observer based field oriented induction motor drives, *Proc. IECON'92*, pp. 67-71.

238. Kubota, H., Matsuse, K., and Nakano, T., "DSP-based speed adaptive flux observer of induction motor," *IEEE Transactions on Industry Applications*, **29**, 2, 1993, pp. 344-348.

239. Li, W., and Venkatesan, R., "A new adaptive control scheme for indirect vector control system," *Conf. Rec. IEEE-IAS'92*, pp. 524-530.

240. Liaw, C.M., Chao, K.H., and Lin, F.J., "A discrete adaptive field-oriented induction motor drive," *IEEE Transactions on Power Electronics*, **7**, 2, 1992, pp. 411-419.

241. Lorenz, R.D., and Lawson, D.B., "A simplified approach to continuous on-line tuning of field oriented induction machine drives," *IEEE Transactions on Industry Applications*, **26**, 3, 1990, pp. 420-424.

242. Lorenz, R.D., and Van Patten, K.W., "High-resolution velocity estimation for all-digital AC servo drives," *IEEE Transactions on Industry Applications*, **27**, 4, 1991, pp. 701-705.

243. Matsuse, K., and Kubota, H., "Deadbeat flux level control of high power saturated induction servo motor using rotor flux observer," *Conf. Rec. IEEE-IAS'91*, pp. 409-414.

244. Moreira, J.C., Hung, K.T., Lipo, T.A., and Lorenz, R.D., "A simple and robust adaptive controller for detuning correction in field oriented induction machines," *IEEE Transactions on Industry Applications*, **28**, 6, 1992, pp. 1359-1366.

245. Nandam, P.K., Cummings, G.F., and Dunford, W.G., "Experimental study of an observer-based shaft sensorless variable speed drive," *Conf. Rec. IEEE-IAS'91*, pp. 392-396.

246. Ohm, D.Y., Khersonsky, I., and Kimzey, J.R., "Rotor time constant adaption method for induction motors using DC link power measurement," *Conf. Rec. IEEE-IAS'89*, pp. 588-593.

247. Ohnishi, K., Ueda, Y., and Miyachi, K., "Model reference adaptive system against rotor resistance variation in induction motor drive," *IEEE Transactions on Industrial Electronics*, **33**, 3, 1986, pp. 217-223.

248. Ohtani, T., Takada, N., and Tanaka, K., "Vector control of induction motors without shaft encoder," *IEEE Transactions on Industry Applications*, **28**, 1, 1992, pp. 157-164.

249. Okuyama, T., Fujimoto, N., Matsui, T., and Kubota, Y., "A high performance speed control scheme of induction motor without speed and voltage sensors," *Conf. Rec. IEEE-IAS'86*, pp. 106-111.

250. Pietrzak-David, M., de Fornel, B., Nogueira Lima, A.M., and Jelassi, K., "Digital control of an induction motor drive by a stochastic estimator and airgap magnetic flux feedback," *IEEE Transactions on Power Electronics*, **7**, 2, 1992, pp. 393-403.

251. Rodriguez de Oliveira, J.C., and de Fornel, B., Microprocessor implementation of an adaptive estimator for induction motors," *Proc. IECON'92*, pp. 127-132.

252. Rowan, T.M., Kerkman, R.J., and Leggate, D., "A simple on-line adaption for indirect field orientation of an induction machine," *IEEE Transactions on Industry Applications*, **27**, 4, 1991, pp. 720-727.

253. Rubin, N.P., Harley, R.G., and Diana, G., "Evaluation of various slip estimation techniques for an induction machine operating under field-oriented control conditions," *IEEE Transactions on Industry Applications*, **28**, 6, 1992, pp. 1367-1375.

254. Saitoh, T., Okuyama, T., and Matsui, T., "An automated secondary resistance identification scheme in vector controlled induction motor drives," *Conf. Rec. IEEE-IAS'89*, pp. 594-600.

255. Schauder, C.D., "Adaptive speed identification for vector control of induction motors without rotational transducers," *Conf. Rec. IEEE-IAS'89*, pp. 493-499.

256. Stephan, J., Bodson, M., and Chiasson, J., "Real-time estimation of the parameters and fluxes of induction motors," *Conf. Rec. IEEE-IAS'92*, pp. 578-585.

257. Sugimoto, H., and Tamai, S., "Secondary resistance identification of an induction motor applied model reference adaptive system and its characteristics," *IEEE Transactions on Industry Applications*, **IA-23**, 2, 1987, pp. 296-303.

258. Sul, S.K., "A novel technique of rotor resistance estimation considering variation of mutual inductance," *IEEE Transactions on Industry Applications*, **25**, 4, 1989, pp. 578-587.

259. Tajima, I., and Hori, Y., "Speed sensorless field-orientation control of the induction machine," *IEEE Transactions on Industry Applications*, **29**, 1, 1993, pp. 175-180.

260. Tamai, S., Sugimoto, H., and Yano, M., "Speed sensorless vector control of induction motor with model reference adaptive system," *Conf. Rec. IEEE-IAS'87*, pp. 189-195.

261. Tsuji, T., Iura, H., and Hirata, A., "Vector approximation with DC link current control and identification in AC drive," *Conf. Rec. IEEE-IAS'92*, pp. 563-569.

262. Vainio, O., Ovaska, S.J., and Pasanen, J.J., "A digital signal processing approach to real-time AC motor modelling," *IEEE Transactions on Industrial Electronics*, **39**, 1, 1992, pp. 36-45.

263. Velez-Reyes, M., Minami, K., and Verghese, G.C., "Recursive speed and parameter estimation for induction machines," *Conf. Rec. IEEE-IAS'89*, pp. 607-611.

264. Verghese, G.C., and Sanders, S.R., "Observers for faster flux estimation in induction machines," *Proc. PESC'85*, pp. 751-760.

265. Wang, C.G., Novotny, D.W., and Lipo, T.A., "An automated rotor time constant measurement system for indirect field-oriented drives," *IEEE Transactions on Industry Applications*, **24**, 1, 1988, pp. 151-159.

266. Yang, G., and Chin, T.H., "Adaptive speed identification scheme for vector controlled speed sensorless inverter-induction motor drive," *Conf. Rec. IEEE-IAS'91*, pp. 404-408.

## Position and Speed Control Systems

267. Awaya, I., Miyake, I., Kato, Y., and Ito, M., "New motion control with inertia identification function using disturbance observer," *Proc. IECON'92*, pp. 77-81.

268. Benzi, F., Zinger, D., and Lipo, T.A., "A new approach to induction motor torque and speed control," *Proc. Conf. on Applied Motion Control CAMC'86*, pp. 25-29.

269. Boldea, I., and Nasar, S.A., "Torque vector control (TVC) - A class of fast and robust torque, speed, and position digital controllers for electric drives," *Electric Machines and Power Systems*, **15**, 1988, pp. 135-148.

270. Boldea, I., and Trica, A., "Torque vector controlled (TVC) voltage-fed induction motor drives - Very low speed performance via sliding mode control," *Proc. ICEM'90*, pp. 1212-1217.

271. Boldea, I., and Nasar, S.A., "Vector Control of AC Drives," *CRC Press*, 1992.

272. Bose, B.K., "Sliding mode control of induction motor," *Conf. Rec. IEEE-IAS'85*, pp. 479-486.

273. Bose, B.K., "Power Electronics and AC Drives," *Prentice-Hall*, 1986.

274. Brickwedde, A., "Microprocessor-based adaptive speed and position control for electrical drives," *IEEE Transactions on Industry Applications*, **IA-21**, 5, 1985, pp. 1154-1161.

275. Cordeschi, G., and Parasiliti, F., "A variable structure approach for

speed control of field-oriented induction motor," *Proc. EPE'91*, pp. 222-226.

276. Enjeti, P.N., Ziogas, P.D., Lindsay, J.F., and Rashid, M.H., "A new PWM speed control system for high-performance AC motor drives," *IEEE Transactions on Industrial Electronics*, **37**, 2, 1990, pp. 143-151.

277. Haneda, H., and Nagao, A., "Digitally controlled optimal position servo of induction motors," *IEEE Transactions on Industrial Electronics*, **36**, 3, 1989, pp. 349-360.

278. Ho, E.Y., and Sen, P.C., "A microcontroller-based induction motor drive system using variable structure strategy with decoupling," *IEEE Transactions on Industrial Electronics*, **37**, 3, 1990, pp. 227-235.

279. Ho, E.Y., and Sen, P.C., "Control dynamics of speed drive systems using sliding mode controllers with integral compensation," *IEEE Transactions on Industry Applications*, **27**, 5, 1991m pp. 883-892.

280. Jezernik, K., Curk, B., and Harnik, J., "Variable structure field oriented control of an induction motor drive," *Proc. EPE'91*, pp. 161-166.

281. Kawamura, A., Ito, H., and Sakamoto, K., "Chattering reduction of disturbance observer based sliding mode control," *Conf. Rec. IEEE-IAS'92*, pp. 490-495.

282. Kubota, H., and Matsuse, K., "Robust field oriented motor drives based on disturbance torque estimation without rotational transducers," *IEEE-IAS'92*, pp. 558-562.

283. Lorenz, R.D., and Yang, S.M., "AC induction servo sizing for motion control applications via loss minimizing real-time flux control," *Conf. Rec. IEEE-IAS'89*, pp. 612-616.

284. Murata, T., Tsuchiya, T., and Takeda, I., "Vector control for induction machine on the application of optimal control theory," *IEEE Transactions on Industrial Electronics*, **37**, 4, 1990, pp. 283-290.

285. Okuyama, T., Fujimoto, N., Matsui, T., and Kubota, Y., "A high performance speed control scheme of induction motor without speed and voltage sensors," *Conf. Rec. IEEE-IAS'86*, pp. 106-111.

286. Park, M.H., and Kim, K.S., "Chattering reduction in the position control of induction motor using the sliding mode," *IEEE Transactions on Power Electronics*, **6**, 3, 1991, pp. 317-325.

287. Park, M.H., and Won, C.Y., "Time optimal control for induction motor servo system," *IEEE Transactions on Power Electronics*, **6**, 2, 1991, pp. 514-524.

288. Prakash, R., Rao, S.V., and Kern, F.J., "Robust control of a CSI-fed induction motor drive system," *IEEE Transactions on Industry Applications*, **IA-23**, 4, 1987, pp. 610-616.

289. Rossi, A., and Tonielli, A., "A unifying approach to the robust control of electrical drives," *Proc. IECON'92*, pp. 95-100.

290. Sabanovic, A., and Bilalovic, F., "Sliding mode control of AC drives," *IEEE Transactions on Industry Applications*, **25**, 1, 1989, pp. 70-74.

291. Sangwongwanich, S., Ishida, M., Okuma, S., Uchikawa, Y., and Iwata, K., "A time optimal single-step velocity response control scheme for field-oriented induction machines considering saturation level," *IEEE Transactions on Power Electronics*, **6**, 1, 1991, pp. 108-117.

292. Stefanovic, V.R., and Nelms, R.M. (Coeditors), "Microprocessor Control of Motor Drives and Power Converters," Tutorial Course, *IEEE-IAS*, 1992.

293. Stojic, M.R., and Vukosavic, S.N., "Design of microprocessor-based system for positioning servomechanism with induction motor," *IEEE Transactions on Industrial Electronics*, **38**, 5, 1991, pp. 369-378.

294. Takahashi, I., and Asakawa, S., "Ultrawide speed control for induction motor covered $10^6$ range," *Conf. Rec. IEEE-IAS'87*, pp. 227-232.

295. Takahashi, I., and Iwata, M., "High resolution position control under 1 sec. of an induction motor with full digitized methods," *Conf. Rec. IEEE-IAS'89*, pp. 632-638.

296. Tzou, Y.Y., and Wu, H.J., "Multimicroprocessor-based robust control of an AC induction servo motor," *IEEE Transactions on Industry Applications*, **26**, 3, 1990, pp. 441-449.

297. Utkin, V.I., "Sliding mode control design principles and applications to electric drives," *IEEE Transactions on Industrial Electronics*, **40**,

1, 1993, pp. 23-36.

298. Vagati, A., and Villata, F., "AC servo system for position control," *Proc. ICEM'84*, pp. 871-874.

299. Won, C.Y., Kim, S.C., and Bose, B.K., "Robust position control of induction motor using fuzzy logic control," *Conf. Rec. IEEE-IAS'92*, pp. 472-481.

300. Zhang, J., and Barton, T.H., "Robust control of an induction machine drive with optimal sliding mode approach," *Proc. IECON'92*, pp. 49-59.

# Index

Adaptive
    control of reference slip speed, 188-189, 209-210, 212-214
    estimation of motor parameters, 187-188
    schemes, 185-189
Airgap flux
    control, 169
    orientation, 168-174
    sensor, 100, 102, 106
Allowable stator current, 61
Axis
    direct, 2
    quadrature, 2
Block diagram
    of DC motor, 90
    of induction motor, 90-92
    of field-oriented induction motor, 94
Braking, 50, 126-127
    dynamic, 127
    regenerative, 127
Current
    control
        in scalar torque control systems, 59
        in current source inverters, 142
        in vector control systems, 107, 162-163, 170
        in voltage source inverters, 138-140
    equation, 15
    reference system
        for airgap flux orientation
            direct, 171
            indirect, 173
        for stator flux orientation
            direct, 164
            indirect, 166
    ripple, 131
DC link, 126-127, 141
DC motor, 64, 88-90
Decoupling
    equations
        for airgap flux orientation, 170
        for stator flux orientation, 162
    system
        for airgap flux orientation, 171
        for stator flux orientation, 163
Equivalent circuit of induction motor
    dynamic T, 20-24, 101

    dynamic Γ, 44-47
    dynamic Γ', 54-56
    steady-state T, 22
Example motor, 30
Fictitious revolving coils, 16-17
Field Orientation Principle, 43, 87-95, 97
    conditions, 93
Flux
    calculator, 101-102, 108, 112-113
    control
        in scalar control systems, 51, 62,
        in vector control systems, 93-94, 104, 107, 161, 164, 169
    controller, 104
Flux-producing current, 57, 107
Freewheeling diodes, 126
Hall sensors, 100, 102
Inverters, 104, 125-157
    current source, 104, 125, 138-144, 147, 155-157
        current control, 142
        frequency control, 142
        phase-angle control, 142
        PWM, 143-144
    harmonic spectra, 131-132, 137
    hysteretic controllers, 138-139
        tolerance bandwidth, 139
    modulation index, 134
    operating modes
        PWM, 132-138, 143-144
        square-wave, 130-132, 141-143
    states, 128-130
        duty ratios, 134-135
    switching
        interval, 133
        variables (signals), 128
    voltage source, 104, 125-138, 145-153
        current control, 138-140
        frequency control, 130
        voltage control, 130-138
    voltage space vectors, 133-134
Model Reference Adaptive System, 179, 185
Mutual inductance
    in T equivalen circuit, 13
    in Γ equivalent circuit, 45
    in Γ' equivalent circuit, 55
Observers, 176-184, 204-209
    rotor flux, 181, 183-184, 207-209
    speed, 177-182, 204-208
        open-loop, 177-179, 204-207

# INDEX

closed-loop, 179-182, 207-208
Phasor diagram of stator currents, 2
Position control, 189-190, 218-223
    linear feedback, 190-191, 218-220
    variable structure, 196, 221-223
Power
    electrical, 47
    input, 71
    efficiency, 71
    factor, 72
    mechanical, 47
    output, 71
Primitive motor
    three-phase, 2
    two-phase, 5
Pulse Width Modulation techniques
    regular sampling, 137
    random, 137-138
    space vector, 132-135
Rotor
    inductance
        in T equivalent circuit, 13
        in $\Gamma$ equivalent circuit, 45
    flux
        control, 104, 107
        orientation, 98-108
    leakage
        factor, 169
        inductance, 13
    resistance
        actual, 12
        in T equivalent circuit, 12
        in $\Gamma$ equivalent circuit, 45
    speed
        of 2-pole motor, 7
        of $P$-pole motor, 50
    time constant, 59
Safe operating area, 60-66, 162
Semiconductor power switches, 126
Scalar control, 43-86, 228
    Constant Volts/Hertz, 43, 47-52
    speed control system, 52-54
    Torque Control, 56-59
    torque control system, 59-66
Shaft
    position sensor, 106
Slip, 21
Slip speed, 7, 50
    control, 59, 188-189
    of 2-pole motor, 7
    of $P$-pole motor, 50
Space vectors, 1-8, 133-134
Speed
    reference

        motor, 52
        slip, 52, 106
        synchronous, 106
    rotor
        of 2-pole motor, 7
        of $P$-pole motor, 50
    synchronous,
        of 2-pole motor, 7
        of $P$-pole motor, 50
Speed control
    linear feedback, 190-191, 214-215
    scalar, 47-54, 77-81
        load changes, 78-80
        reversing, 80, 81
        starting, 77-79
    vector, 189-190, 214-218
        linear feedback, 214-215
        variable structure, 216-218
Stator
    flux
        control, 161, 164
        orientation, 160-168, 197-201
    inductance
        in T equivalent circuit, 13
        in $\Gamma$ equivalent circuit, 55
    leakage inductance, 13
    magnetomotive forces, 2
    resistance, 12
Supply frequency, 2
Torque
    angle, 95
    calculator, 103, 113-114
    constant, 103, 108
    controller, 104
    developed, 47
    estimator, 189
    harmonic, 148
    peak, 50
    per flux squared limitation
        with airgap flux orientation, 170
        with stator flux orientation, 163
    program, 117-123
    starting, 49
Torque control
    scalar, 56-66, 82-86
        at reversing, 84-86
        at starting, 82-84
    vector, 93-94, 104, 107
Torque equation
    in excitation reference frame, 30, 91
    in stator reference frame
        for T equivalent circuit, 19, 20
        for $\Gamma$ equivalent circuit, 46
        for $\Gamma$ equivalent circuit, 56

INDEX 255

of field-oriented induction motor
    with airgap flux orientation, 168
    with rotor flux orientation, 93
    with stator flux orientation, 160
    steady-state
        for T equivalent circuit, 25
        for $\Gamma$ equivalent circuit, 47
        for $\Gamma'$ equivalent circuit, 56
Torque-producing current, 57, 107
Torque production
    in DC motor, 88-90
    optimal conditions, 88-90
Total leakage factor, 161
Transformations, direct and inverse
    rotor to stator, 8
    stator reference frame to excitation reference frame, 28, 99
    T equivalent circuit to $\Gamma$ equivalent circuit, 44, 45
    T equivalent circuit to $\Gamma'$ equivalent circuit, 54, 55
    three-phase to stator reference frame, 9, 10
Turns ratio, 8
Universal field orientation, 173
Variable Structure Control, 192-196
    chattering, 194, 216
    hysteretic relay controller, 194
    reaching condition, 192
    relay control, 192-193
    sliding mode, 192
    sliding surface, 192
    switched-gains control, 192-193
    switching function, 192
    tolerance band, 192
Vectors, see Space vectors
Vector control systems, 159-223
    with airgap flux orientation, 168-174, 199-201
        direct, 170-171, 200-201
        indirect, 171-173, 199-201
    with current source inverters, 175-176, 202-205
    with rotor flux orientation, 98-123
        direct, 100-105, 117-118
        indirect, 106-108, 118-123
    with stator flux orientation, 160-168, 197-201
        direct, 164-166, 200-201
        indirect, 166-168, 197-203
    with universal field orientation, 173
Voltage control
    in scalar speed control systems, 51, 52
    in voltage source inverters, 130-138
Voltage equation
    in excitation reference frame, 29
    in stator reference frame
        for T equivalent circuit, 13-15
        for $\Gamma$ equivalent circuit, 46
        for $\Gamma'$ equivalent circuit, 55
    steady-state, 25
        for T equivalent circuit, 25
        for $\Gamma$ equivalent circuit, 46
        for $\Gamma'$ equivalent circuit, 56